TEXAS WOOLLYBACKS

TEXAS WOOLLYBACKS

The Range Sheep and Goat Industry

by

PAUL H. CARLSON

Texas A&M University Press
COLLEGE STATION

Library of Congress Cataloging in Publication Data

Carlson, Paul Howard.
Texas woollybacks.

Bibliography: p.
Includes index.
1. Sheep industry—Texas—History. 2. Goat industry—Texas—History. 3. Wool trade and industry—Texas—History. I. Title.
HD9436.U53T483 1982 338.1'763145 82-40311
ISBN - 13: 978-0-89096-133-9 (cloth)
ISBN - 13: 978-1-62349-418-6 (paper)
ISBN - 13: 978-1-62349-507-7 (ebook)

Manufactured in the United States of America
Second printing, 2016

For my parents
Howard and Alpha Carlson

Contents

List of Illustrations

MAPS

List of Tables

Preface

THE sheep and goat industry in Texas is an old and honorable business. Although it lacks some of the majestic qualities of the cattle and oil industries, it has nevertheless been an economically and socially important enterprise that has proven to be a staple business in Texas since the time Spaniards first settled the state.

Because for many years they considered woollybacks (a popular name for sheep in the 19th century) ignoble, silly, and smelly, Anglo Texans left sheep raising to immigrants, including Englishmen, Basques, Mexicans, Germans, and others. But when it became apparent that wool growing was profitable, farmers and ranchers did not hesitate to enter the business. A boom followed, and wool growing spread rapidly to all parts of the state. Livestockmen soon discovered that there is some truth in the Texas adage: "You raise cattle for prestige, you raise sheep for money."

Work on this book started several years ago when R. Sylvan Dunn, former director of the Southwest Collection at Texas Tech University, suggested that I investigate the sheep and wool industry. I found the subject interesting and largely unexplored. My study produced several articles published in regional histories and a short manuscript for the Food and Fiber National Institute of Achievement. When I decided to broaden my research to include the entire scope of the sheep and goat industry of Texas, my project met with the enthusiastic encouragement of Seymour V. Connor and Ernest Wallace, both of Texas Tech, who felt that there was need for such a work.

My aim in this book, therefore, has been to prepare a history of the range sheep and goat industry in Texas from early times to about 1930, after which most woollybacks were confined within mesh-wire fences. I have tried to provide a fresh synthesis in a straightforward narrative, with analysis of the more important phases of the state's sheep and goat industry.

Two small research grants from Texas Lutheran College and a larger one from the Food and Fiber National Institute of Achievement provided important financial aid. Editors of the *Panhandle-Plains Historical Review* and the *West Texas Historical Association Year Book* have graciously allowed me to include material, in chapters 6 and 8, respectively, that appeared in somewhat different form in their journals.

The material of many libraries and archives contributed to the substance of this work. To the staffs of the libraries at Texas Lutheran College, Texas Tech University, Southwest Texas State University, and the Barker History Center at the University of Texas I owe a debt of gratitude. In particular I would like to acknowledge the valuable assistance of Byron Price of the Panhandle-Plains Historical Museum, whose sympathetic advice played no small role in the preparation of this book, and David B. Gracy and his staff at the archives of the Texas State Library.

Several other individuals provided assistance. In early stages of the study A. J. Taylor shared with me her notes and a manuscript she had written on sheep in the Panhandle. She provided important bibliographical information. Bill Sims, executive director of the Texas Sheep and Goat Raisers Association, offered help and encouragement. Hiram Phillips and Elmer Kelton of San Angelo gave me information and arranged several interviews. I owe thanks to Michael Hooks, Janet Brown, and E. Karl Alf who helped to check some sources at museums and libraries I could not reach. I alone accept responsibility for whatever errors may exist.

Most of all I am obliged to my good friend David J. Murrah, university archivist and director of the Southwest Collection at Texas Tech University. Without his valuable assistance this volume could not have been written. My wife Ellen and our children proved themselves patient and understanding. Their self-sacrifice during research and writing extended sometimes beyond reason.

TEXAS WOOLLYBACKS

1
The Spanish Period

SPAIN dominated Texas and much of the Western Hemisphere for three hundred years. During that long period she spread her economic, cultural, and political institutions over a vast area, establishing powerful roots that remain to the present day. Her contributions are legion, but in few places are they more substantial and direct than in the Texas livestock industry, which owes its methods, terms, and paraphernalia to the Iberian nation that held Texas against all challengers until 1821.

The range sheep and goat industry in Texas began with the daring and visionary Christopher Columbus. On November 3, 1493, Columbus, during the second of his four voyages to the New World, unloaded sheep, goats, and other livestock on the island of Hispaniola. These animals formed the beginning of a colorful and expansive livestock industry. The start proved a slow one, however, as only a few others besides Columbus seemed interested in establishing a thriving agricultural community. Most colonists wanted only to search for gold or other mineral treasures of the Caribbean.

Nevertheless, the livestock industry grew. Bands of goats were introduced from the Canary Islands, and colonists continued to bring with them domesticated sheep and other animals from Spain. In addition, Columbus, governor of all lands he discovered, urged in a 1496 letter to King Ferdinand that licenses to search for gold in the New World be limited to certain seasons as a precaution against the neglect of livestock.[1] Although Ferdinand apparently took little action on the letter's recommendation, Columbus's interest in agriculture helped to

[1]See C. W. Towne and E. N. Wentworth, *Shepherd's Empire*, pp. 5, 15–25, 120–21.

promote sheep and goat raising. As time passed Spaniards took up residence throughout the Caribbean region, and then in 1521 colonists planning to settle on the Isthmus of Panama, led by Gregorio Villalobos, brought sheep to the mainland. During the same year, Hernando Cortes's sudden conquest of the highly civilized and gold-rich Aztec Indians attracted many more people to New Spain (Mexico). Spaniards followed Cortes swiftly and in large numbers. To promote livestock raising, a royal condition was imposed upon persons who planned to settle New Spain requiring that "within a time specified there must be from ten to thirty settlers, each with one horse, ten milch cows, four oxen, one brood mare, one sow, and twenty ewes. . . ."[2]

Domestic animal populations increased. As Spaniards augmented the sheep and goat populations and other species of domesticated livestock with additional, but not required, importations from Spain, the animals spread throughout the islands of the West Indies and over extensive portions of the mainland. Cortes himself imported large numbers of sheep to his ranch near Cuernavaca. That the sheep and goats adapted quickly to their new habitat was shown by the extraordinary numbers of them later observed grazing on the hillsides and plateaus of South America. By 1555, sheep had become common and cheap.

Meanwhile, sheep and goats had entered North America. Antonio Mendoza, the first viceroy of New Spain and a skilled Spanish empire builder, and his loyal follower, the well-born and wealthy rancher-politician Francisco Vásquez de Coronado, were responsible. Viceroy Mendoza encouraged thousands of Spaniards to emigrate, asking them when possible to bring livestock. Many of them did. The European animals, especially cattle, pigs, horses, sheep, and goats, increased so rapidly that meat was abundant, and semi-wild roaming herds became a danger to life and property, trampling fields and stampeding through villages.

In 1540 Mendoza appointed Francisco de Coronado to lead a grand expedition northward from Mexico City to the upper reaches of the Rio Grande to find the mythical seven cities of Cibola, whose fabled streets were paved with gold and whose walls were studded with turquoise, and the equally wealthy kingdom of Quivira. The subse-

[2]Quoted in J. M. Jones, "History of the Range Sheep Industry in Texas," *Southwestern Sheep and Goat Raiser* 6 (March, 1936): 8.

quent Coronado expedition was a large one indeed. He took with him cavaliers, foot soldiers, Indians, Franciscan missionaries, and herds of cattle, horses, sheep, and other animals. For two years he directed his men on explorations that covered much of the American Southwest. Of the sheep, many were eaten, and some were left along the way or lost to predators. However, the expedition spared enough sheep to provide small breeding flocks for Fray Juan de Padilla and Fray Luis de Escolona, the Franciscans who chose, when Coronado returned to New Spain in 1542, to stay behind with Indians. Within six months Indians had killed both priests, and the sheep had probably met the same fate.

Several years passed before sheep appeared again. Then, during the 1580s interest in New Mexico, as the upper Rio Grande region was called, flared anew. The English sea hawk and pirate Sir Francis Drake may have stimulated this interest. Drake, after sailing through the Straits of Magellan, emerged in the Pacific, where he raided Spanish towns and *galeones* believed safe from the English. He sailed northward to California before turning westward to circumnavigate the globe. Horrified by Drake's accomplishment, Spain believed that he had found the fabled sea route through America—the Strait of Anian or Northwest Passage.

So, perhaps partly in response to Drake's stunning action, in 1581 Father Agustín Rodríguez and a dozen companions, seeking converts for the Spanish church and a northern stronghold for the Spanish empire, visited Indian pueblos as far north as Taos. Sheep accompanied them. The next year a merchant-adventurer named Antonio de Espejo, with a friar and fourteen soldiers, led an independent expedition up the Rio Grande to Taos. He, too, carried sheep. Encouraged by the economic possibilities of New Mexico, Espejo issued an enthusiastic report, which inspired a flurry of proposals by various noblemen who wanted to conquer the northern province. But, in part because King Philip II had determined to invade England in 1588 (a disastrous event that resulted in the destruction of the powerful Spanish Armada), it was not until 1597 that the Spanish crown finally approved the colonization project of Don Juan de Oñate.

Oñate, a rich mineowner from Zacatecas, got his expedition underway in 1598. As governor, he had charge of 130 soldier-colonists, ten Franciscan friars, and a large number of servants and Indians. He took with him eighty-three wagons and 7,000 head of livestock, including some 2,900 sheep and 716 goats. The animals helped to feed and clothe the colonists at San Juan, the village Oñate established on the Rio Grande north of present Santa Fe. Although Oñate wanted to search for gold, his men preferred ranching, farming, or Christian missionary work. As the years passed, his colonists and others who drifted in laid out new towns, such as Santa Fe, erected mission stations, and found that sheep raising was the key to their prosperity. Sheep "took naturally to the sparse grasses of that semiarid region," writes one historian, "multiplying so rapidly that within a few years as many as 15,000 yearly were driven southward to Chihuahua and another 25,000 to Viscaya where they were in such demand that they were used as a medium for exchange."[3]

In time the upper Rio Grande sheep industry spread. It pushed out of the valley and onto the high tableland to the east and into low mountain ranges to the west. Some early herders, *pastores* they were called, took sheep into the upper Pecos River drainage system. From there the industry expanded north and south along breaks of the famous stream.

Then, suddenly, unrelated developments brought sheep and goats into Texas. Perhaps the most important events were the activities occurring after René Robert Cavelier, Sieur de La Salle, an aspiring French empire builder, landed in 1685 at Matagorda Bay on the Texas Gulf coast. Word of the intrusion, reaching Mexico in the form of exaggerated tales from Indians, caused panic in Mexico City. Spanish officialdom saw La Salle's settlement as a threat against New Spain that needed to be removed. There followed several expeditions to oust La Salle and place Spanish outposts north of the Rio Grande in Texas.

In one such expedition in 1689 Alonso de León, governor of Coahuila, located the ruins of La Salle's settlement, destroyed by Indians two years before, and the charred bones of its murdered defenders. He had with him cattle and horses, some of which were purposely released along major streams to propagate. Herbert Eugene Bolton's translations of the reports do not specify whether De León brought sheep and goats.[4]

[3] Ray Allen Billington, *Westward Expansion*, p. 356.

[4] Herbert Eugene Bolton, ed., *Spanish Exploration in the Southwest, 1542–1706*, pp. 388–423.

In another expedition the next year De León and Fray Damián Massanet, a Franciscan priest, with a party of 110 soldiers and four missionaries crossed Texas to the fertile Neches River Valley. Here they established among the Tejas natives a mission station, San Francisco de los Tejas. The tiny outpost served to guard Texas from further French intrusions and to convert the Indians.

In a third episode, Don Domingo Terán de los Ríos organized a sizeable *entrada*. In addition to horses and cattle, Terán in 1691 brought to Texas some 1,700 sheep and goats. His small livestock, the sheep and goats, posed problems at the Rio Grande, which, swollen by heavy spring rains, flowed swift and wide. Nevertheless, when his men on horseback carried the sheep and goats over the river, a safe crossing was accomplished. Later, as the party moved overland toward the Neches River, hot, dry weather slowed the expedition. Near the end of July it halted for a time to allow the flock to catch up. Through the next month intense heat continued to slow the Spaniards, but they reached their destination in East Texas with the sheep and goats in good health.

Two years later, because of a fierce Indian uprising, the Spanish left East Texas. Presumably their sheep and goats perished or left with them. Although two decades followed during which Texas remained unoccupied, when Spanish authorities established the presidio of San Juan Bautista in 1699 and three Catholic missions across from present Eagle Pass, the frontier of northeastern Mexico advanced to the Rio Grande.

There it hovered for a time, but in 1716, as French economic and political activity in Louisiana increased, Spain set out again to occupy east Texas. In charge was Captain Don Domingo Ramón, in command of twenty-five soldiers, nine missionaries, and an amazing French trader named Louis de St. Denis. St. Denis, by showing up at San Juan Bautista in 1714, had initiated the Ramón *entrada*. Ramón took along horses, mules, and oxen as well as a flock of one thousand sheep and goats. Twelve animals drowned when trying to wade the Rio Grande, but Ramón managed to get his livestock across the other Texas rivers without great difficulty. At the Brazos, Ramón, according to Carlos E. Castañeda, arranged for sixty Indians to take "the goats across, one by one, for which [service Ramón ordered] two goats and an ox killed for them." Reaching the Neches River in July, his men fell to rebuilding the mission of Nuestro Padre San Francisco de los Tejas and others, including the mission San Miguel de Linares de los Adaes, only fifteen miles from the Red River, and the mission Nuestra Señora de los Dolores de los Ais, between the Neches and Sabine rivers about forty-five miles east of Nacogdoches. Six in all were built, each well equipped and guarded by a small presidio. Here, Ramón's sheep and goats served as food for the missionaries, soldiers, and Indian converts.[5]

Quickly other people followed Ramón. One of these was Martín de Alarcón. As governor of Coahuila and Texas, Alarcón is credited with establishing in 1718 both the presidio and the village of San Antonio de Béxar near the mission San Antonio de Valero, later known as the Alamo, which had been founded a few days earlier by Fray Olivares. The fort was to guard the mission and the route between East Texas and Mexico. When he entered Texas, Alarcón brought with him 548 horses, six droves of mules, at least sixty beeves, and an unspecified number of hogs, chickens, sheep, and goats. Castañeda reports that Alarcón, in a letter to the viceroy of New Spain dated September, 1717, indicated that he had accumulated two hundred cattle and one thousand sheep for the expedition.[6] Two years after San Antonio was created, Spaniards established a mission named Concepción a few miles away. Later they erected others nearby. Bands of sheep grazed at each mission.

In 1720 the Marqués de San Miguel de Aguayo, the wealthy soldier-governor of Coahuila, to strengthen Spanish fortifications on the Louisiana frontier, led a large force into Texas. Well-provisioned with a year's supply of flour, corn, and livestock, the five hundred mounted troops were organized into eight companies. Each group was responsible for 350 horses, six hundred cattle, and eight hundred sheep. Severely troubled by drought first and heavy rains and cold weather later, the expedition nevertheless brought approximately three hundred cattle and four hundred sheep through from Nuevo León to Los Adaes Mission near Natchitoches in present-day Louisiana—a distance of about nine hundred miles. After completing his work and leaving behind a well-garrisoned presidio on the East Texas frontier, Aguayo returned home during the winter of 1721–22, pausing only to build a station and a fort at Matagorda Bay. Over the following two decades Spain quietly consolidated her position.

[5] Carlos E. Castañeda, *Our Catholic Heritage in Texas, 1519–1936*, II, 53.
[6] Ibid., II, 45–55, 84, 91.

Then, beginning about 1740, Spain's northern outposts in Texas and elsewhere were threatened. England had founded Georgia and threatened Spanish supremacy in Florida. The French from Canada and Louisiana were edging into Texas and across the Great Plains toward New Mexico. A number of Indian tribes menaced positions across the north. In response, from about 1740 to 1765, Spain pressed additional settlements and outposts, planning to tame and hold Texas as a defensive province. New missions were created in the San Antonio area, and by 1745 all of them possessed functioning irrigation ditches and raised fruits and vegetables in such abundance that surpluses were sold to presidio soldiers. Related to the San Antonio establishments was a presidio on the Guadalupe River near modern Victoria, known as Nuestra Señora de Loreto, or La Bahía, with an accompanying mission called Espíritu Santo de Zuñiga. These had been founded near Lavaca Bay in 1717, but in 1726 they were moved inland and in 1749 were relocated at the San Antonio River near present Goliad. Flocks of sheep and goats grazed in the area.

In 1746 Captain José de Escandón, Conde de Sierra Gorda, aided by Franciscan missionaries, began the conquest of the region from Lavaca Bay along the coast and southward through the lower Rio Grande to the Pánuco River in Mexico. After a complete exploration of the region was undertaken, Escandón recruited ranchers and others in Mexico to form a civilian rather than a military colony. "In 1748," writes Sandra L. Myres, "nearly four thousand prospective settlers gathered at Querétaro and marched northward, driving great herds of horses, cattle, burros, sheep and goats."[7] The caravan resembled those American pioneers later took to Oregon and California. The colony, Nuevo Santandar in Mexico, developed into a center for stock-raising activity that spread northward across the Rio Grande. Escandón's efforts resulted in the founding of twelve new villas, or towns, south of the Rio Grande and others in Texas to the north, such as Villa de Balmaceda on the lower San Antonio River and the Villa de Vedoya at the mouth of the Nueces. The latter proved unsuccessful.

Failure at the mouth of the Nueces stimulated further settlement on both sides of the river in the Rio Grande Valley. In 1755 Captain Tomás Sánchez founded Laredo, which became an important Rio

[7] Sandra L. Myres, *The Ranch in Spanish Texas*, p. 15.

Grande crossing on the highway from Saltillo to San Antonio, ultimately displacing San Juan Bautista. Much of the lower valley filled up with a thin string of *haciendas* ("ranches"), but not settlements, extending in every direction. Since the district turned immediately to livestock, thousands of sheep, goats, and cattle grazed in the region.

Encouraged by success in the lower Rio Grande and spurred by French intrusion in the east, Comanche pressure in the north, and Apache pleas for assistance, Spain founded additional missions, presidios, and civil colonies. These include one along the lower Trinity, some on the San Marcos, Llano, San Gabriel, and San Saba rivers, and a few in other scattered and isolated places. Since Spain soon abandoned most of them, they served the government only briefly.

As a result of the military and missionary activity, by 1750 sheep and goats had become well established in Texas. The type or kind of these sheep, however, is a matter of dispute. Most authorities, including Edward N. Wentworth, former director of Armour Livestock Bureau in Chicago and author of several studies on sheep, maintain that the animals were descendants of the *chaurro*, an inferior race of sheep in Spain. Lean and gaunt, from the modern viewpoint they were distinctly scrubs weighing only sixty to eighty pounds at maturity and yielding from one to two and a half pounds of course, open wool. According to Wentworth, their descendants retrogressed until they became the modern Navajo sheep, with which many Southwest sheepmen are familiar. Many of them developed multiple horns, and some of them were so goatlike that people frequently considered them goat-sheep hybrids. The Spanish sheep had to resist northers, escape coyotes, mountain lions, or wolves, find their own feed and water, and in general preserve a preponderance of traits having little to do with good carcass or fleece values. Under such conditions, stock degeneration was unavoidable.[8] Settlers in Texas considered them more important as mutton than as wool animals.

Likewise, the goats brought to Texas are believed to have been but the common stock of Spain. Primarily meat animals, they were long-legged and small of body, gave some milk, and provided other essential dietary elements, such as cheese, to their owners. The meat,

[8]Edward N. Wentworth, "Sheep Trails of Early Texas," *Southwestern Sheep and Goat Raiser* 9 (June, 1939): 7.

when properly prepared, was similar to venison. Tough and lean, the goats could walk great distances, would eat almost anything, and required little attention. Since Spanish herders preferred *cabrito* ("meat of the young goat") to mutton, goats nearly always accompanied sheep drives. Later authorities claimed three types of Spanish goats had been included. One variety had short, coarse hair that was bright and glossy. Another had short hair with a dark, kempy appearance. The third—a descendant of the Maltese goat—grew long, straight hair of sometimes five or six inches.

The widely separated mission stations at which the sheep and goats grazed were unique Spanish frontier institutions. They not only provided an outlet for the crusading zeal of the padres, but also helped secure the frontier. Near each station Indian villages were established at which the missionaries taught Christianity and handicraft skills while the natives provided a labor force for the agricultural and pastoral pursuits associated with each complex. Hoping to make the Indians and the missions self-supporting as soon as possible, the priests maintained close supervision and tight discipline when training the neophytes in the workshops and gardens or on the farms and ranches. Sometimes enterprising colonists, such as a group of Canary Islanders who in 1732 settled near missions at San Antonio with five hundred sheep they had received from the Spanish king, took up residence nearby, creating civil communities far in advance of the slowly northward-moving permanent settlements.

Some of the missions were elaborate. Many grazed thousands of cattle, horses, sheep, and goats, of course, but some enjoyed irrigated, fenced-in fields of hundreds of acres, where in season the priests and their Indian laborers grew corn, beans, melons, peaches, potatoes, and other items. Some missions had granaries, a carpenter shop, an iron shop, a furnace for burning lime, a tailor shop, and a workshop where very good cotton and woolen cloth was woven. The cloth was primarily for home consumption. Since most of the Spaniards preferred beef to mutton, the sheep and goats remained secondary enterprises to cattle and corn, and wool production was always small. Few of the goat flocks were large. Goats, left to themselves much of the time, were used for milk and meat. Large, regular markets for surplus production were so distant and so difficult to reach as to be virtually nonexistent.

There was little structure and overall organization to the livestock industry. With the population small, experienced labor scarce, and no regular markets, goats and cattle sometimes scattered untended on the open range. One historian writes that "when the settlements or missions needed meat or decided to make a shipment of hides or tallow, riders went out to capture the necessary animals and bring them in for branding or slaughter."[9] As a result, during the mid-eighteenth century herds of unbranded livestock, especially the hardy cattle and horses, roamed the isolated plains and valleys near the mission fields and settlements.

In most of the older mission fields the animals thrived. In fact, due to the Spanish practice of breeding the sheep twice yearly, their numbers increased rapidly. By 1727, for example, sheep in the San Antonio area had become the chief means of mission support. Missionary fathers with their Indians and nearby settlers grazed approximately 9,000 sheep in the vicinity. As the years passed the number of sheep continued to increase, with several authorities setting the count in 1765 at nearly 12,000 sheep and goats. Two years later Father Gaspar José de Solís, inspector of missions for the College of Zacatecas, checked the missions at San Antonio. He reported that Spaniards pastured 5,487 cattle, more than 600 saddle horses, almost 1,000 breeding mares, more than 100 donkeys, and 17,000 sheep and goats in the region.[10]

At La Bahía livestock also flourished. De Solís reported about 2,200 sheep at Mission Espíritu Santo de Zuñiga and Nuestra Señora del Rosario. Because mission fathers in the area had difficulty raising crops for lack of rain and irrigation networks, they annually traded cattle and other livestock to the San Antonio missions for a supply of corn.[11]

The lower Rio Grande enjoyed an even larger livestock population. Although many of the missions and settlements there, such as those at San Juan Bautista, Mier, San Bernardo, and Reinosa (Reynosa), were located south of the Rio Grande, thousands of sheep and

[9] Myres, *Ranch in Spanish Texas*, p. 16. See also Herbert Eugene Bolton, *Texas in the Middle Eighteenth Century*, pp. 58–59.

[10] P. P. Forrestal, trans., *The Solís Diary of 1767*, pp. 5–7, 20, 36. See also Castañeda, *Catholic Heritage in Texas*, IV, 5–15, 36, 265–66; Val W. Lehmann, *Forgotten Legions, Sheep in the Rio Grande Plain of Texas*, pp. 11–15.

[11] K. S. O'Connor, *The Presidio la Bahía del Espíritu Santo de Zuniga, 1721 to 1846*, pp. 32, 42, 48.

goats grazed on *haciendas* in Texas, north of the river. Some authorities estimate that in 1757 near the sites of present McAllen and Pharr some 13,000 sheep and goats as well as 3,000 horses and 1,200 head of cattle pastured. In the same year at Rancho de Dolores, about twenty miles below Laredo, nearly 8,000 horses, cattle, and mules, and twice as many sheep were found. Near Laredo, Captain Sánchez had only a few cattle but pastured about 10,000 horses, mules, and sheep.[12]

The East Texas woodlands, however, proved a disappointment for Spanish sheep and goat raisers. Although the few Spanish settlers in the area introduced thousands of sheep, mission reports in the 1760s showed few animals of any kind in the region. In 1767 Nacogdoches claimed only "eighty sheep and asses, thirty oxen, fifty cattle, twenty-five gentle horses, twenty mules, and two droves of mares." Mission de los Ais had in 1768 only fifteen to twenty mules, about ten cows, a few bulls, oxen, and saddle horses.[13]

In the newer mission fields sheep and goat populations remained small. In 1757, for example, Governor Jacinto de Barrios y Jáuregui attempted unsuccessfully to establish fifty families from Saltillo with more than nine thousand sheep and other livestock along the lower Trinity. Here among the Orcoquiza Indians Spaniards built a presidio and planted crops, but no settlement was established. The mission and presidio at San Saba lasted only a short time, suffering all the while from Indian attacks and lack of civilian support. Other new sites proved similarly unsuitable and subject to Indian attack.

Ranching north of the Rio Grande, on the other hand, gradually increased and extended to the Nueces River. In 1766 Escandón reported to the viceroy that "the settlement [of the country] between the Rio Del Norte [Rio Grande] and the Nueces, which you . . . charge me with, is making much progress, and I hope that it will be most useful in the colony." Blas María de la Garza Falcón, founder and captain of Camargo, had a ranch near present-day Petronilla, below the Nueces River with "a number" of people and "a stock of cattle, sheep and goats, and cornfields."[14] Also, some ranches began to appear between San Antonio and La Bahía.

[12]Castañeda, Catholic Heritage in Texas, IV, 5–15, 36, 265–66.

[13]Forrestal, *Solís Diary*, p. 20; Castañeda, *Catholic Heritage*, IV, 5–15, 36, 265–66; Lehmann, *Forgotten Legions*, p. 15.

[14]Cited in Bolton, *Texas in the Middle Eighteenth Century*, p. 301.

The few Spaniards who came to the scattered ranches, following in the footsteps of the *conquistadores* and missionaries, often hoped to acquire vast holdings in Texas and then use Indian labor to develop their *haciendas.* That had been the practice elsewhere in the Spanish world. They quickly learned, however, that the situation was quite different in Texas: Indians in the province could be neither conquered nor converted. The lordly Comanches, the truculent Apaches, and the fierce tribes of the coast, including Karankawas, Tonkawas, and others, refused to submit tamely to a life of farming or ranching for someone else. Since crops grew well in the bountiful climate of South and East Texas, and cattle and horses ran wild in the province, most early Spanish Texans became farmers and stock raisers. Corn was the staple crop, but inhabitants also produced beans, chilis, and a little sugarcane. Cattle became the dominant livestock, although sheep, goats, horses, pigs, and others grazed, too.

As for sheep and goats, the more prosperous ranchers hired servants from among the poorer class to tend the animals. Customarily servants and masters entered into contracts as to the length of service and wages. A common practice had the servant-sheepherder tending bands of about fifteen hundred sheep. He remained with the animals day and night, moved them from one range to another, and participated in lambing, shearing, and other activities associated with sheep raising. There was little trade in wool, and the Spaniards used the animals for meat. As time passed, sheep production became secondary to cattle raising.

By 1765 colonization of Texas had begun to slip, population to decline, and political authority to weaken. As a result Visitor-General José de Gálvez received instructions from King Charles III, one of the ablest men of the Spanish Bourbon line, to reexamine Spanish policies in Texas and across the north. De Gálvez sent the indefatigable Marqués de Rubí on a tour of inspection. After visiting San Saba, San Antonio, Nacogdoches, Los Ais, Los Adaes, La Bahía, and other outposts in Texas during the summer of 1767, the highly intelligent and keenly observant Rubí recommended that Spain quit most of the region and use the southern fringe as a defensive border. A royal decree in 1772 followed his ideas. It called for the abandonment of all missions and presidios except San Antonio de Béxar and La Bahía, the strengthening of San Antonio by moving to it settlers from Los Ais and Los Adaes, and a new Indian policy.

Teodoro de Croix, a Frenchman, as commander of the Spanish Internal Provinces, accepted the task of putting the Rubí recommendations into force. Assisting him was Deputy Governor of Spanish Louisiana Athanase de Mézières, a Frenchman who had settled in Natchitoches. After completing most of the work, De Mézières visited the region to report to the Spanish viceroy in Mexico about conditions in East Texas. In 1776 he noted that disgruntled civilians from Los Adaes and Los Ais had refused to take up residence in San Antonio. Under their leader, Antonio Gil Ybarbo, they had established Nuestra Señora del Pilar de Bucareli, a settlement on the Trinity River, where they had planned to raise cattle, horses, sheep, and goats and to produce tallow, lard, and soap. But since the site was subject to repeated flooding and Comanche raids, the unhappy migrants had returned to an old mission site near Nacogdoches. The settlement remained small, numbering about 380 residents in 1783, and concentrated on Indian trade and cattle and horse raising for a livelihood.

De Mézières also indicated in 1776 that there were approximately 300 sheep and goats near Natchitoches. The flocks were owned by Frenchmen, he said, rather than by Spanish settlers, but, as most of them bore Spanish brands, they probably had been stolen from the Spanish regions farther west. According to the census taken at the beginning of the year, he wrote, "There is stock to the number of 1,258 head of horses, 1,842 cattle, over 300 sheep and goats, and 782 hogs." Three years later he reported on the price of livestock in San Antonio: a fat cow, four pesos; a three-year-old sheep, six pesos; a breeding ewe, three pesos; half-broken horses, six pesos per head; mares in droves, one peso per head; and wild mules, eight pesos a head. Edward Wentworth notes that the relative values of the different classes of livestock attest to the large number of cattle and horses in the Texas province. The relatively high price for wild mules probably resulted from the facts that they were less numerous and that the Spaniards in Texas had begun to use them widely. Since a three-year-old sheep was worth more than a fat cow and as much as a half-broken horse, one might conclude that the numbers of sheep in the territory had declined.[15]

In addition to the problems attendant on a declining population, livestock interests in Texas faced another difficulty—taxation. In 1780

[15] Herbert Eugene Bolton, ed., *Athanase de Mézières and the Louisiana-Texas Frontier, 1768–80*, II, 120, 241. See also Edward N. Wentworth, *America's Sheep Trails*, p. 31.

De Croix issued an order declaring all unbranded stock the property of the Spanish crown and subject to public claim. Anyone could acquire unmarked animals merely by obtaining a capture permit and paying a tax equal to about fifty cents for each animal taken. According to Carlos Castañeda in his monumental work on the Catholic background of Texas, authorities within the Spanish church quickly responded. Bishop Rafael José Verges, clerical head of Nuevo León, Nuevo Santander, and Texas, proclaimed that tithes were due on captured animals and on the permits issued for their taking.[16]

Although De Croix's order did not become law until some years later, its immediate effect was a massive and frenzied stock hunt in which ranchers, missionaries, soldiers, and Indians participated. The resulting roundup was so large that by 1788 taxes had been paid on nearly 24,000 animals. De Croix's order did not destroy the herds, flocks, or the missions, but its long-range effect deprived the missions at La Bahía and San Antonio of a major source of income and contributed to the population decline. Although a new mission built at Refugio in 1791 had little difficulty in procuring sufficient stock and meat to supply its needs, the Spanish mission system continued its decline, and in 1794 Governor Manuel Muñoz announced that all missions would be secularized. The missionaries delayed the process when they could, but by 1824 all Texas missions had closed their doors. As they closed, the padres distributed some of their livestock, including sheep and goats, to the mission Indians. Other livestock, especially cattle and horses, wandered off, becoming part of the many herds of wild, unbranded stock roaming the woods and prairies.

From De Croix's order in 1780 until Mexican independence in 1821, Texas became less and less stable. Notwithstanding all of Spain's efforts to people the area, the population dwindled until Spain had only the settlements at San Antonio, La Bahía, and Nacogdoches. Between these scattered outposts there were vast stretches of land occupied only by wild game and roving bands of Indians. No towns existed on the Texas side of the Rio Grande above Laredo. From the mouth of the Rio Grande north and east along the entire Gulf shoreline to Matagorda Bay, there was not a single Spanish community. The population of the huge Texas area, consequently, was small in 1780—

[16]Castañeda, *Catholic Heritage in Texas*, V, 1, 27–32, 177–88.

perhaps four thousand persons, more than one thousand of whom were soldiers.[17]

Similarly, the sheep and goat populations entered a period of decline. The confusion after Indian raids that left some pastures bare of stock, the withdrawal of herders to below the Rio Grande, and the wars for Mexican independence cut the size of flocks. Natural mortality also was heavy, especially during droughts, as in the 1780s, or in times of unusually severe winters, as in 1786. Whatever the exact causes for the decline, a livestock population of at least 40,000 to 50,000 sheep and goats, carried into Texas during the Spanish mission period and before, had by 1800 largely disappeared.

There still remained a few sheep and goats at San Antonio, La Bahía, and near the Gulf coast below the mouth of the Nueces. But as markets for them withered, the number of animals dwindled. For a brief time a small sheep and goat trade between Texas and Missouri, the Ohio Valley, and Tennessee developed, but that broke down after 1800 when Spain ceded the Louisiana Territory back to France. Following the purchase of Louisiana by the United States in 1803, the trade was never reestablished legally. Spanish authorities in Mexico City feared the Americans even more than the ousted French. Accordingly, they placed embargoes and restrictions on trade, effectively stopping it. Shortly afterward, the Spanish period ended.

Clearly, Spain made several lasting contributions to the range sheep and goat industry. She had shown that the animals, even when left unattended, could thrive in the area. In extreme South Texas her practice of breeding sheep twice a year would be followed for almost another one hundred years. Spanish stock, methods, terminology, and organization continued in use long after Spain left the province. Although Spaniards continued to play important roles in the industry, such political developments as Mexican independence turned their attention temporarily away from economic and agricultural pursuits. And, as the Spanish focused on other matters in the early nineteenth century, Americans, Englishmen, and Germans dominated the Texas range sheep and goat industry.

[17] LeRoy R. Hafen, W. Eugene Hollon, Carl Coke Rister, *Western America*, p. 201; Odie B. Faulk, *A Successful Failure*, p. 162.

2
The Early American Period

THE period between the decline of Spanish influence in the 1790s and the retreat of Mexico from the region in the 1840s witnessed the birth of Anglo-American interests in the Texas range sheep and goat industry. Developments during the time encouraged expansion of sheep husbandry, produced efforts to improve sheep breeds, attracted immigration of European herders, and wrought other changes. Major alterations, indeed, they began as United States citizens entered Texas and continued through the Republic era and afterward.

Anglo-Americans entered Texas around the turn of the nineteenth century. The first who reached the area from the east and northeast were traders, such as Philip Nolan, who with a handful of men searched for wild horses and cattle, or filibusters, such as Augustus Magee, who with a small military force invaded the area unsuccessfully. The intruders looked upon the region as a land of easy opportunity, whose rich river bottoms beckoned farmers and whose large Indian populations attracted traders. In the three decades after 1790 more than a dozen men led hundreds of others on trading, hunting, exploring, and military forays into the area.

While the penetration of Texas by traders and adventurers at the turn of the century weakened Spain's grip on her far northern frontier, revolution in New Spain cut the European political hold altogether. Begun in 1810 by Father Miguel Hidalgo y Costilla, a soldier-priest of Dolores, the struggle for Mexican independence enjoyed only limited success until thirty-three-year-old Agustín de Iturbide, a Creole officer in the Spanish army, in 1820 joined the rebel cause. Victory came quickly afterward, and by August, 1821, Mexico was free of Spain.

In the meantime, Spain had adopted a new plan to control and develop Texas. She had hoped to hold the region by admitting Anglo-

Americans as permanent settlers. But before she could carry through her plans, the Mexicans secured their independence. The new regime in Mexico City thereupon concluded arrangements in 1823 for granting a huge tract of land to Stephen F. Austin, with the understanding that he would bring to Texas three hundred families. The Austin deal locked open the American door to Texas. As other grants followed, aggressive, pioneer Americans poured into the Mexican state, attracted by a favorable climate, a chance to "start over," and a liberal land policy that allowed an individual to purchase a *sitio*—4,338 acres—for two hundred dollars or less. The Anglo population of Texas increased rapidly, and within seven years there were more than 20,000 Americans in Texas, with contracts outstanding for thousands of additional people. The effect was to remove much of the land area of Texas from control of the Mexican government and place it in the hands of land agents, or empresarios, like Stephen F. Austin, Green C. DeWitt, Sterling C. Robertson, James McGloin, James Power, James Hewetson, and John McMullen.

From the beginning of Anglo-American penetration, settlers agreed that cotton and corn production represented the best possibilities for immediate economic advancement. Since most of the early settlers were southerners, coming out of Louisiana, Missouri, Tennessee, and Alabama, it was understandable that cotton and corn received the most attention. Prior to independence farmers grew little cotton near San Antonio. But as early as 1828 in the region just to the east, which included Stephen Austin's grant, they produced five hundred bales and doubled that amount the next year. Substantial increases occurred in the following years.

From the present Houston area eastward to the Sabine River, cattle and horse raising were the primary pursuits. Ranchers introduced improved breeds and drove their animals to market at New Orleans, a long trip made more difficult by many river crossings. Taylor White, sometimes called the first cattle king of Texas, used slave labor for herding on his huge Trinity River spread, which included 40,000 acres of land and held perhaps 30,000 head of cattle.

The number of sheep and goats in Texas is indeterminable, but certainly it was less than the figures reported in the Spanish years of the middle eighteenth century. There had been hundreds of flocks on the Texas side of the Rio Grande, but by the turn of the century most of the sheep had disappeared above the river. In the East Texas border country Frenchmen pastured a few hundred sheep and goats. Between these extended points the timid creatures could be found in large numbers only near San Antonio and La Bahía, and there probably not exceeding five thousand head.[1]

Nevertheless, Stephen Austin and his partners were aware of possibilities of Texas sheep raising. As they built their towns, plantations, or farms, they wrote of the soil, climate, and presence of Mexican sheep, suggesting that American stock could upgrade the Spanish *chaurros*. Indeed, according to Edward N. Wentworth, the first move to improve the blood of the Spanish sheep in Texas had occurred as early as 1821 through the efforts of pioneer farmers near present San Felipe. The colonists brought in sheep from the northeastern states and bred them to the Spanish stock with good results. The quality and quantity of production far exceeded those of the Mexican flocks. Instead of two-pound fleeces, noted Wentworth, six to eight pounds of wool were shorn, and instead of lambs producing eighteen-pound carcasses at slaughter and wethers producing twenty-five- to thirty-pound carcasses, there were fat sheep yielding carcasses that averaged thirty-five to forty pounds.[2]

Such results brought attention. In early 1822 Thompson H. Ficklin, a textile entrepreneur from Caledonia, Missouri, wrote to Austin that he had seen and read some of Austin's letters describing Texas. "It struck me," Ficklin suggested, "that a man could in a short time accumulate considerable wealth by the establishment of a clothes factory in your country." He indicated that he had almost ready to go into operation a fulling mill, a carding machine for wool and cotton, and two spinning machines, one for wool and one for cotton. "And owing to the great quantity of sheep," he continued, "and the cheapness of wool in your country," assuming there were also streams or rivers sufficient to drive machinery without too much expense, "I am of the opinion such an establishment would not only be of an advantage to the inhabitants but would be profitable to the owner."[3]

Austin's reply, if he made one, is no longer extant. The young em-

[1]John J. Linn, *Reminiscences of Fifty Years in Texas*, p. 98.

[2]Edward N. Wentworth, "Sheep Trails of Early Texas," *Southwestern Sheep and Goat Raiser* 9 (July, 1939): 23.

[3]Th. H. Ficklin to Stephen F. Austin, January 8, 1822, in Eugene C. Barker, ed., *The Austin Papers*, I, 462–63.

presario, however, could have indicated, as a prospectus for Texas published in Cincinnati later did, that factories could be established at the falls of the San Antonio, Guadalupe, Colorado, and Brazos rivers, all within 250 to 300 miles of the Rio Grande, where wool sold for as low as three to four cents a pound.[4] He might also have pointed out that should farmers of the Brazos-Colorado-rivers region turn their attention to the growing of wool, the long haul from the Rio Grande would no longer be necessary.

There were enough animals in Texas that Austin was concerned about the registry of livestock. In February, 1824, he wrote to San Antonio to his friend Baron F. E. N. de Bastrop, the commissioner empowered to grant Mexican land titles, who had helped to secure Austin's grant. Austin indicated that the interest of the colony required that a recorder's office be established to register marks and brands of hogs, cattle, sheep, and other livestock. "You know," Austin argued, "that in the United States there is a recorders office of this kind in every county and the inhabitants here are therefore accustomed to it and will not be satisfied unless one is properly established."[5] Two years later Austin had a recorder's office.

In the next few years the population of Texas increased rapidly as empresarios moved to settle their land grants. Austin, DeWitt, and others wrote favorably of their Mexican state. In 1829 Austin penned a note to Commodore David Porter, an American merchant who planned to establish a colony south of the Rio Grande, urging him to consider Texas. The climate, Austin admitted, did not allow cultivation of coffee or cacao or tropical productions, but there were other advantages. The pasturage, he said, "or 'range' as we term it, is certainly superior to any thing I have seen in any country, and the facilities for raising cattle, horses, mules, sheep, and hogs, etc. almost exceeds credibility." Within a year Porter had applied for a grant in Texas. Others agreed with Austin, prompting nearly every empresario to encourage his settlers to mix sheep and goats with the horses and cattle. One colonizing agent wrote that "never was there a country better calculated for sheep than that of Texas."[6]

[4] Cited in Wentworth, "Sheep Trails of Early Texas," pt. 2, p. 23.

[5] Austin to Baron de Bastrop, February 3, 1824, in Barker, *Austin Papers*, I, 738.

[6] Austin to Commodore David Porter, February 16, 1829, in Barker, *Austin Papers*, II, 166; Wentworth, "Sheep Trails of Early Texas," pt. 2, p. 23.

The business was not always easy, however, for unfenced sheep needed watching. In the frontier country, if coyotes or wolves did not get the sheep, Indians might. Reports in the Mirabeau Lamar papers noted that in the 1830s Comanche as well as other Indians, including Karankawas on the coast, Tonkawas in the interior, and Wacos and Tawakonis along the Brazos River above the San Antonio Road, regularly stole small numbers of sheep. Sometimes the Indians were caught, as happened once on the McMullen-McGloin grant. John McMullen and James McGloin, natives of Ireland but residents of Matamoros, in 1828 had established a colony of Europeans and Americans along the San Patricio River, and in 1830 settlers at San Patricio were visited by several parties of Comanches. One night in mid-October a warrior got into a sheep pen, wrote one observer, and "was about taking one when he was Seized by the neck by a Servant of McGloin and gave him a floging [*sic*]. . . ."[7] Other sheep wandered off, getting lost in the brush or falling victim to predators.

Despite such hazards, the sheep and goat populations increased, if slowly. Immigrant pioneers in Texas brought sheep with them. A few Mexican herders reoccupied the Gulf coast south of the Nueces, and flocks along the lower Rio Grande increased again. Colonel Juan Almonte, a Mexican soldier and statesman commissioned to proceed to Texas and report his observations to the president of Mexico, wrote mistakenly in 1830 that no sheep grazed in the Nacogdoches district. In fact, a few farmers and some plantation owners had them in pens, but their numbers were small. Almonte also claimed that no sheep grazed in the Brazos district, the area around Gonzales and Harrisburg, including Austin and DeWitt's grants. In this he was essentially correct. There was little grazing of sheep, but several farmers had them in pens. Finally, Almonte said that in the San Antonio de Béxar district sheep were scarce, not exceeding five thousand head.[8]

Farther south, however, there were large numbers of sheep and goats. John J. Linn observed in his reminiscences that on a trip in the 1830s from Victoria to Matamoras he encountered about fifty miles from the Rio Grande a Mexican herder, his wife, and three children, who possessed three thousand sheep, some goats, twelve dogs, and

[7] C. A. Gulick, Katherine Elliott, and Harriet Smither, eds., *The Papers of Mirabeau Bonaparte Lamar*, V, 379. In the same collection (III, 515), see also R. Lambe, Matagorda, Texas, to M. B. Lamar, April 24, 1841.

[8] Quoted in Linn, *Reminiscences*, pp. 98–102.

four horses. There were many animals in the area around the mouth of the Nueces River, where in the 1750s Blas María de la Garza Falcón had established his Rancho de Santa Petronilla with twenty families and a large number of cattle, horses, sheep, and goats. One authority has claimed that in 1835 three million head of livestock grazed on the Texas-Mexican frontier but that the number of sheep and goats did not exceed twenty thousand head.[9]

Then, suddenly, Texas' brief but successful struggle for independence from Mexico brought changes to the sheep and goat industry. More people immigrated to the new Republic of Texas, bringing with them additional livestock. The war's end also marked a slow, steady change in the method of handling sheep. On the substantially self-sufficient farms of the early settlers, the pasture was an adjunct to the field, and sheep were kept in pens close to the family house by night and sent to a nearby pasture each morning. During the republic years the pasture became more important than the field, and a ranch system of production developed, replacing the earlier farm system. Another change wrought by the short struggle for independence found farmers, herders, and others trailing sheep northward out of the Mexican states of Tamaulipas, Coahuila, and Nuevo León, where tens of thousands of the animals grazed. From these states sheepmen drove flocks into the Texas counties west of the Brazos and below the San Felipe de Austin–San Antonio region. By 1840 competition had filled the market, and according to one account large amounts of specie from the sheep trade had entered San Antonio.[10]

Sheep raisers improved the Mexican breed. Some of them bred Spanish Merinos to the *chaurros*. The Merino, a broad-backed, smooth-bodied, large-carcassed animal, was developed in Spain. Its ability to walk long distances in compact flocks was one reason this fine-wooled sheep, rather than British mutton sheep, became the range sheep of Texas and the Southwest. The Spanish Merino, after it first came to America about 1810, revolutionized the wool industry of the United States and started America on the way to becoming one of the world's greatest wool producers.

[9] Ibid., p. 16; J. M. Nance, *After San Jacinto: The Texas-Mexican Frontier, 1836–1841*, p. 8.

[10] *Houston Telegraph and Texas Register*, February 3, 1841, as quoted in Nance, *After San Jacinto*, p. 416.

Not everyone used Merinos. Robert Lambe, who came from England to settle two *sitios* of land (nearly 9,000 acres) thirty miles above San Patricio on the Nueces River, imported his own livestock, including fine English Southdown sheep.[11] The Southdown, a large, smooth-bodied, medium-fleeced animal, was one of the favorites across the entire South because it would breed and lamb quickly, thus bringing early returns to the owner. But hardier grazers soon pushed it out of Texas and the Southwest.

Another early favorite in Texas was the English long-wooled Cheviot. Raisers liked it for its fattening qualities and its sturdy constitution in resisting cold and hunger. But, under Texas conditions of the 1830s and 1840s, Cheviots did not crossbreed well in the effort to improve sheep stock. When they were bred to Southdowns, according to Lamont Johnson, a sheep-breeding expert, the quality of the original fleece was reduced.[12] There were other favorites, too, most of which are no longer bred in the United States. Improved sheep, whatever the variety, entered Texas south and east of a line defined roughly by the present-day cities of Marshall, Corsicana, Waco, Austin, Boerne, Uvalde, and Eagle Pass.

In addition to Americans and Englishmen, there were others who brought in purebreds. Scotsmen settled around San Antonio, and Germans pushed into the region of New Braunfels, Boerne, and Fredericksburg. In addition, according to Edward Wentworth, "a plentiful supply of immigrants from Kentucky, Tennessee, Ohio, Michigan, and Illinois was sprinkled among the foreign settlers."[13] While most newcomers introduced their own particular kind of sheep, the commercial stock of the state remained the *chaurro* and its relatives, which provided the foundation of all production.

Efforts of the German settlers were particularly significant. Compared with total population, the numbers of Germans settling in Texas were impressive. By the time Texas achieved its independence slightly more than 100 German families had come to the region, mostly in the area between the lower Brazos and Colorado rivers. Between 1836 and 1845, when Texas became a state, some 13,000 or more Germans en-

[11] Robert Lambe to M. B. Lamar, April 24, 1841, in Gulick, Elliott, and Smither, *Papers of Lamar*, III, 515.

[12] Lamont Johnson, "Sheep Breeds 100 Years Ago," *Sheep and Goat Raiser* 27 (June, 1947): 29.

tered the republic. From then until the outbreak of the Civil War another 12,000 to 15,000 Germans arrived. Since most of them were yeoman farmers, they were accustomed to handling livestock, including sheep.[14]

As they scattered over Texas, the German immigrants made serious efforts to establish a good sheep breed. Ottoman von Behr, author, farmer, and rancher, who came to Texas in 1846, was among the first Germans to upbreed the animals. Settling at Sisterdale in present Kendall County, he began building a flock. He bred two imported German Merino (Saxon) rams with a flock of *chaurros* he had purchased at one dollar a head. By 1854 he had a small but growing flock of quality sheep grazing unpenned on his pastures.

Saxon sheep, developed in Germany in the eighteenth century from a Merino strain, were large, fine-wooled animals. Herders of the king of Saxony, using careful methods of selection, managed the original flock. At one time a royal edict had ordered that "every farmer must use rams from the royal herd or be evicted from the land." The decree served the dual purpose of improving the Saxon sheep and ensuring a market for the rams produced on the royal farm. So great was the development that Saxon rams were sold the world over. Many went to Australia when it was rising to leadership in the production of wool and mutton. Many others came to America in the 1840s with German immigrants.

Another successful young German, H. J. Richarz, settled in 1849 on a five-hundred-acre sheep ranch near San Antonio. Using several Saxon rams he had brought with him from Europe, he built up a small flock. Four years later he moved to D'Hanis in Medina County, where he established a ranch on which he raised cattle and sheep. In 1854 Frederick Law Olmsted, who traveled through Texas for the New York *Times*, described Richarz's sheep operation as consisting of 300 good animals produced by breeding the Mexican ewe to Saxon rams. Richarz was also among the first Texas ranchers to provide hay and shelter for his animals during the winter months.[15]

[13] Edward N. Wentworth, *America's Sheep Trails*, p. 379.

[14] See Bobby Weaver, "German Contributions to the Texas Cattle Industry," typescript, 1977, photocopy in Southwest Collection, Texas Tech University, pp. 2–4. See also Terry G. Jordan, *German Seed in Texas Soil: Immigrant Farmers in Nineteenth-Century Texas*, pp. 40–45, 54.

Using men like von Behr and Richarz as examples, one observer in 1849 explained that in Texas wool showed the greatest potential for profit of any livestock product. Some wool growers, the man reported, showed returns of over 33 percent per year. He claimed that Texas required only one-tenth the number of herders needed to tend the same number of livestock in northern regions of the United States. Moreover, he reported that Mexican sheep could be purchased for twenty-five to fifty cents each, considerably less than the dollar a head von Behr had paid for his *chaurro* ewes.[16]

The potential profits attracted many people. They also caused a few hardy and daring entrepreneurs to drive improved sheep to Texas from the Ohio Valley. Francis M. Pool, a South Carolinian who had emigrated westward in the early 1850s, was a pioneer sheep trailer. Pool covered on horseback much of the Texas sheep country, ranging from Fredericksburg to Kyle, Austin, and Waco. From San Antonio in 1856–57, he went to Illinois, where he bought a flock of sheep. Driving them with the help of an assistant and two dogs, he came southwest across the Mississippi River into Missouri and thence into northwestern Arkansas. After wintering in Arkansas, he trailed the sheep southward to Kyle, Texas, where he sold them, but not before he had marketed their wool clip in New Orleans. Although not registered animals, the sheep were heavier-boned and heavier-fleeced than the native *chaurros*, and they carried some mutton blood.[17]

Perhaps 50,000 sheep from the Midwest and the Ohio Valley entered Texas during the 1850s. D. A. Nichols led one of the last drives before the Civil War. His equipment included a covered wagon and a team, a riding horse, a portable stove, water kegs, a small tent, dogs, guns with ammunition for hunting and protection, and assorted food supplies. Nichols purchased his sheep in Illinois and entered Missouri at Hannibal, having ferried the animals for one-half cent per head, the standard price, across the Mississippi. His route proceeded past Boonville and Carthage before entering Indian Territory, where he turned south for the Red River and Texas. At streams with toll bridges he paid

[15] Frederick Law Olmsted, *A Journey Through Texas*, p. 259.

[16] Arthur L. Finck, Jr., trans., "The Regulated Emigration of the German Proletariat with Special Reference to Texas, by Dr. Von Herff" (Master's thesis, University of Texas, Austin, 1949), pp. 68–69; Linn, *Reminiscences*, pp. 98–102.

[17] Wentworth, "Sheep Trails of Early Texas," p. 34.

the customary fee of five dollars for flocks of 2,000 head or more. He waded or swam the animals across minor streams as well as the Canadian and Red rivers. As he traveled, Nichols penned his sheep each night as a protection from wolves. In Illinois and Missouri he rented space at convenient farms, but beyond Carthage and through Indian Territory he fashioned from yard-wide sheeting portable pens mounted on stakes. The pens were designed to prevent sheep from crowding in the corners and could be rolled for hauling. Some sheep drivers used pens of rope and stakes.[18]

Nichols, like the majority of drivers, went only part way the first season, wintering in Jasper County, Missouri. He planned to acclimate the flock to the warmer temperatures and to avoid the weight shrinkage and death losses that took place when the fall trip was made in one season. It cost most drivers, in addition to fitting out the party, about ten cents per head to drive from Illinois to Texas in the 1850s. The cost was met with the sale of the sheep or by sale of the wool sheared from the flock.

Pool and Nichols were speculators who bought sheep in the Midwest and after great overland drives to the new county sold the animals. Other sheepmen, generally those from New England and New York, drove flocks to establish on their own farms. John M. Stephens of Vermont was such an individual. In 1852 he took a small flock of Merinos by steamboat to Mississippi. But because of the hot, humid weather and because of encouragement from his brother, who said Texas was excellent sheep country, he moved to Texas. He settled in 1859 at a point just east of Waco, where he cultivated cotton and raised sheep, hogs, and cattle.[19]

The sheepmen represented part of a rapidly growing Texas population. In 1846 the frontier line of settlement extended from Corpus Christi to San Antonio on the southwest and thence northward roughly past New Braunfels, Fredericksburg, Austin, Waco, and Dallas to Preston on the Red River. Including blacks but not Indians, the population had reached about 142,000 people. The census of 1850 showed about

[18] Wentworth, *America's Sheep Trails*, pp. 78–79. For a poignant, first-hand account of a typical overland drive from the Midwest, see William Edgar Hughes, *The Journal of a Grandfather*, pp. 39–63.

[19] Wentworth, *America's Sheep Trails*, p. 77; Edward N. Wentworth, "The Golden Fleece in Texas," *Sheep and Goat Raiser* 24 (December, 1943): 16.

212,500. Ten years later it had jumped to 604,000. The frontier line of settlement had moved westward to a line running from Henrietta on the north through Belknap, Palo Pinto, Brownwood, Llano, and Kerrville, to Uvalde. The advance halted here temporarily because of the Civil War, hostile Plains Indians, and a lack of timber and water. Along the Rio Grande were located a number of communities, including Laredo and Eagle Pass. East of the mid-1850s frontier line of settlement three-fourths of the population had been born outside Texas.

In the mid-1850s sheep east of the frontier line, totaling about 100,000 head, grazed everywhere. Near Matagorda Bay, Thomas Decrow raised the animals. As an experiment, in 1842 he acquired from a Corpus Christi–area rancher thirteen *chaurro* ewes and bred to them a Bakewell (Leicestershire or Leicester) ram he had purchased in New Orleans. Bakewells were large, long-wooled animals better known for their mutton than for the coarse wool they provided. Decrow found that his sheep stayed healthy and his flock rapidly increased in size. Since there were few wolves in the region, he never used a herder, but let his sheep run undisturbed until there were a substantial number of lambs. Then he gathered the lambs in pens, marked or branded them, castrated the males, and let them go.[20]

In 1844 Decrow bought in New York two Southdown rams. He recommended Southdowns as the best sheep for the prairies and climate of the coast because, "they are good breeders, often have twins, and, being good nurses, raise [their young] well." In subsequent years he added only 1 ewe, but for a few years he allowed Captain George Grimes, an old friend, to graze 16 ewes and 12 wethers on his property. By 1849 Grimes's sheep had increased in numbers to nearly 160, of which Decrow kept one-third, while Grimes removed the others. By 1854 Decrow had more than 1,200 animals.[21]

Not far away on the Brazos River in Fort Bend County, Thomas F. McKinney experimented with sheep. He was, like Decrow, one of the first Texans to attempt to improve the quality of the wool by the introduction of foreign sheep. In 1840 he imported Bakewells from England, Merinos from Vermont, and a cross of Merinos and Southdowns from Ohio. These he bred to *chaurros*. McKinney found the Bakewell unsatisfactory. It did not suit the humid climate, being too fat and sub-

[20] *Texas Almanac for 1859*, pp. 128–29.
[21] Ibid., p. 128.

ject to diseases. He claimed that Bakewells did not breed well. The Mexican sheep died in severe weather but were very prolific. The other varieties had difficulty adjusting to the hog-wallow prairie and rank, coarse sedge grass common in that part of the state. Thus it was that McKinney, unsuccessful in Fort Bend County, moved in 1851 to a beautiful place on Onion Creek near Austin. Here he had uninterrupted good fortune with his sheep, although he gave them little attention. The size of his flocks increased, and the amount of wool produced from each animal nearly doubled. Such results brought attention. McKinney developed a close correspondence and friendship with George Wilkins Kendall, a remarkable wool grower near New Braunfels, and editors of the *Texas Almanac* contacted him for information on sheep and wool production in the state.[22]

In Nueces County beginning in 1856 James Bryden and others grazed sheep at New Santa Gertrudes. Bryden's flock in that year consisted of 694 ewes and 11 Southdown bucks. But the flock increased rapidly. Writing in the *Texas Almanac*, Bryden reported that by 1860 he had 3,778 head, although he had sold or lost 800. He had also sold three clips of wool and was preparing to sell his fourth.[23]

Near Huntsville, Sam Houston, the irrepressible hero of Texas independence and erstwhile president of the Republic of Texas, considered the sheep business. In an 1858 letter to Dr. Ashbel Smith, a wool grower at Galveston Bay and former Texas minister to England and France, Houston wrote, "I have resolved to go into the *sheperdizing* [*sic*] business." He planned to raise both mutton and wool sheep, indicating that graded animals could be purchased in Louisiana and East Texas for $2.50 per head. He preferred, however, to buy from Dr. Smith. "If you will part with your sheep," Houston wrote to his friend, "I will send you a Draft on Galveston, or Houston, for your flock at the rate of *four dollars* per head."[24] Although it is not known where he secured the animals, Houston obtained enough sheep for his farm to start in the business in a limited way. Neighbors who also ran sheep provided him with advice and assistance, although he had help of his own who were experienced in wool production.

[22] Ibid., pp. 124–25.

[23] James Bryden, "Sheep Raising in Nueces County," *Texas Almanac for 1861*, p. 171.

[24] Sam Houston to Ashbel Smith, October 29, 1858, in Amelia W. Williams and E. C. Barker, eds., *The Writings of Sam Houston*, VII, 189.

Southwest of Huntsville in Gonzales County, J. D. Fly, living on the east side of Peach Creek, attracted statewide recognition for his financial success with sheep. In 1855 he bought 150 head, paying $375 for the animals. Each year the size of the flock nearly doubled, and the wool clips likewise improved in size. Within three years the wool clips had paid the original cost of the sheep, and, because the size of his flock had increased, as had the price and the public demand for sheep, his flock's worth was estimated at $1,800. Moreover, he had, according to the San Antonio *Daily Ledger*, which publicized his work, sold mutton worth $5,000.[25]

William Bollaert, who kept a diary of his tour through Texas, noted that plenty of sheep could be found along the Colorado River between Austin and Columbus. On the plantation of a "Mr. H——" there could be found crops of cotton, corn, sugar, tobacco, wheat, and vegetables. Flocks of sheep and goats on the plantation were in "thriving condition," wrote Bollaert. Significantly, the plantation slaves operated a spinning jenny and a loom to make both cotton and wool cloth.[26]

Such household manufacturing was not uncommon in Texas in the early American period. Many farmers with small sheep operations sheared, scoured, spun, and wove their own wool. Household wool-cloth manufacturing was a time-consuming task that went through several processes. Early wool-growing farmers in Texas, especially those from New England and the upper South, accepted the idea that sheep must be washed to clean the wool before shearing. They selected a date during the first warm spring days for the task. When the day arrived, they plunged their animals one by one, into a nearby stream or pond or beneath a waterfall, vigorously scrubbing each for a minute or two before releasing it to a small pen to dry. Although soaps and combs were used, the performance constituted more a ritual than a thorough cleansing measure.

After washing them, the owners sheared their sheep. To shear sheep with a hand clipper quickly, smoothly, and without cutting the animal required skill attained only through long experience. Some owners performed the task themselves, but most hired expert itinerant

[25] Cited in John Ashton, "The Start of Sheep Breeding in Texas," *Sheep and Goat Raiser* 25 (December, 1944): 37.

[26] William Bollaert, *William Bollaert's Texas*, p. 187.

workmen or neighbors who had developed a special proficiency in the art. The standard charge was the equivalent of three to four cents per head, but perhaps a few experts earned more. The shearers, working on *chaurros* and *chaurro* crosses, which had little wool on their bellies, legs, or faces, completed each animal quickly.

After clipping, the wool was again cleaned of grease and dirt—a practice called scouring. In early Texas, women performed the task using the family washtub, homemade soap, and weak lye. After washing it, the women dried the fleece and picked it by hand into a soft and fluffy mass ready for carding. Carding consisted—as it does even today, when the process is performed by machine—of forming the wool into cylindrical rolls, with the wool fibers combed out to lie lengthwise in the roll and parallel to each other so that the fibers can be drawn and twisted into yarn by a rapidly revolving spindle. Women used hand cards, simple implements similar to wire-studded cards used in grooming cattle. After carding, women spun the wool into yarn and turned the yarn into stockings, mittens, or cloth for shirts, dresses, blankets, or other garments.

While many pioneer families were clothed by such unceasing industry, woolen mills appeared well before the Civil War. At Huntsville the state penitentiary, which had opened in 1850, operated a woolen and cotton factory. A report published in 1857 indicated that a building for the engine and boilers was complete, and one for the factory nearly finished. Including expenses for machinery, engine, fixtures, and delivery of the material to Huntsville, cost of the enterprise totaled over $40,000. Estimated expenses for operating the factory came to nearly $10,000 a year.[27] The plant annually manufactured about a million and a half yards of cloth. During the Civil War it distributed much of the cloth to soldiers and their families. The remainder went to customers on the open market. During the Civil War, Texas created a military board to promote manufacture of wool and cotton. The board directed the penitentiary operation. It also established at Austin a foundry for cannon and munitions, which it later converted to the manufacture of plows and spinning jennies. To promote homespun manufacture, the board imported from Europe 40,000 pairs of cotton and wool cards.[28]

But factories appeared. Cotton textile manufacturing was the

[27] *Texas Almanac*, 1857, p. 99.

[28] Ernest Wallace, *Texas in Turmoil*, pp. 127–28.

principal industrial interest in early Texas, and some of the cotton mills could handle wool. In 1860 twenty-four Texans called themselves woolen manufacturers, but only two woolen goods establishments met census definition of a factory. The Houston City Mills Manufacturing Company with 2,288 spindles and one jack of 240 spindles for wool was the largest in the state in the 1860s. By 1870 woolen goods were also made in Walker and Comal counties. Wool-carding and cloth-dressing establishments appeared in Hunt and Grayson counties.[29]

While sheep and wool provided good profits, experience taught East Texas sheepmen that they must get away from the Texas coast and out of the Blackland Belt if they expected to succeed in sheep raising. William Bollaert's Mr. H——, admitting that his sheep along the Colorado were in good condition, insisted that the mesquite "grasses and clear streams of the West [were] better than the part of the Colorado he resided on."[30] Thus, in the early American period sheep moved overland across eastern portions of the country toward the frontier, where they challenged, but did not supplant, cattle raising. Also in the early American period sheep began to move out of the state to distant markets. Although most went by land, some went by ship, leaving from Indianola or Corpus Christi. In the 1850s a few men delivered live sheep by steamboat down the Red River from Jefferson, paying eighty cents per head for the transportation.

As the Texas sheep industry spread, some farmers and ranchers turned to goats. A few raised the common, but hardy, Mexican goat, a popular milk and meat animal of medium size and of mixed origin. Most stockmen turned to the Angora. A native of Turkey (Asia Minor), the Angora has spiral horns and a silky white fleece that hangs down in curly locks all over its body. Although most people prefer the meat of Angoras to that of other species, the Angora is most important for its hair. It makes a sleek cloth called mohair that is widely used for robes, capes, suits, and plush upholstery. Mohair goats, as we know them to-

[29] U.S. Department of Interior, *Eighth Census. Population of the United States in 1860*, p. 491; Vera Dugas, "Texas Industry, 1860–1880," *Southwestern Historical Quarterly* 59 (October, 1955): 151–83. To encourage manufacturing companies, the state provided liberal land grants to individuals who established woolen and cotton mills. Textile operators established two such land-grant mills in New Braunfels, one each in Waco, Hempstead, and Bastrop. All of them handled more cotton than wool (Thomas Lloyd Miller, *The Public Lands of Texas, 1519–1970*, pp. 92–93).

[30] Bollaert, *Bollaert's Texas*, p. 187.

day, are crossbreeds between the delicate Angora and the Spanish goat—a cross that combines good mohair production with hardiness. Angora goats thrive best where there is a variety of brush and shrubbery, which should include live oak and other evergreen browse to afford plenty of variety throughout the year. The vast Edwards Plateau region of Texas offers the best Angora goat browse to be found in the United States and possibly in the world.

While there are conflicting reports on who brought the first Angoras to America, most authorities give credit to Dr. James B. Davis. A physician and a cotton expert of Columbia, South Carolina, Davis, a grand-looking man with large black eyes, black hair, florid complexion, and high white forehead, in 1849 brought a small flock of the goats to the United States. In 1844 in response to a request from the sultan, President James K. Polk had appointed Davis to introduce cotton culture to Turkey. Upon completing his work, Davis returned to the United States, bringing with him several Asiatic animals, including the Angora goats. After experimenting with them a few years, in 1854 Davis sold the purebred Angoras to Colonel Richard Peters of Atlanta, Georgia, who had been watching the work with no little interest from the beginning. Many of the kids produced through his breeding efforts, Davis disposed of for good profit.[31]

There is also conflicting information as to when the first Angoras arrived in Texas. F. O. Landrum, who lived on the Nueces River in Uvalde County, may have been the first to introduce them. However, John Ashton has suggested that Colonel Robert A. Williamson of Gallatin, Tennessee, an agent for the Sumner Cashmere Company, had first brought eight head of Angora to Austin in 1857.[32] Another account has Colonel W. W. Haupt bringing the first Angoras to the state. Haupt, a powerful man who boasted that he never lived in town a day

[31] "History of Mohair," typescript, Mohair Council of America, Records, Southwest Collection, Texas Tech University; "The Report of the Association," printed report of the American Angora Goat Breeders Association, n.d., American Angora Goat Breeders Association, Records, Southwest Collection, Texas Tech University; George T. Willingmyre et al., *The Angora Goat and Mohair Industry*, Misc. Circular No. 50, United States Department of Agriculture, 1929, pp. 13–20.

[32] H. S. Hunter, "El Paso Pool and a Bit of Angora History," *Sheep and Goat Raisers' Magazine* 8 (March, 1928): 11; John Ashton, "The Golden Fleece of Early Days," *Sheep and Goat Raiser* 24 (December, 1943): 15; Jno. A. Black, "The Cashmere Shawl Goat," *Texas Almanac*, 1860, pp. 197–98.

in his life, in 1853 purchased a farm near Kyle in Hays County, where he ran the first steam cotton gin used in his area—Western Texas, as it was called then. In 1858 he determined to herd goats and began raising some of the common stock. Quickly, however, he made arrangements with Colonel Peters to purchase for one hundred dollars eight head of Angoras. The transaction marked the start of the "Haupt goats," which in later years became popular with Texas goat men.[33] Although others, including Joseph P. Devine of San Antonio, who purchased a large number of Angoras from Peters, soon entered the goat business, Angora and mohair production for some time remained a secondary enterprise. Angora numbers in Texas before the Civil War were so small as to be indeterminable.

In the 1850s at the end of the early American period of the Texas range sheep and goat industry, an old Austin-colony settler could look back upon more than three decades of agricultural development. Many changes had occurred. The population of Texas had increased thirtyfold. The frontier line of settlement, even in face of Indian raids and Mexican armies, had been pushed west of the modern cities of Dallas, Waco, Austin, and San Antonio. Land values, after a collapse in the 1840s that left many people land poor, had started upward again. Livestock populations, especially cattle, had multiplied. The eastern portions of Texas had been transformed from a vast wilderness to a settled, prosperous state.

The old Austin-colony pioneer could also see that much remained the same. Cotton and corn remained the state's basic source of revenue and sustenance, while other crops were grown on a small scale. Transportation, or lack of it, for agricultural commodities continued to plague farmers and ranchers. Although slaves worked the fields, crops, and herds of plantations and large farms in East Texas and elsewhere, a large portion of the Texas population remained non–slave-owning yeoman farmers who had developed a high degree of self-sufficiency. Herders, especially in extreme South Texas, the Houston area, and the region of the lower Trinity, still preferred cattle and horses over sheep and goats. Although wool production was beginning to take on some importance, sheep continued to be raised, as they had in the Spanish

[33] Ervin Hickman, "Texas Has Twenty-five Million Dollar Angora Goat Industry," *Sheep and Goat Raiser* 16 (July, 1946): 20.

period, primarily as mutton animals. Goat populations also remained small with the exception of the area below the Nueces, where they provided meat for Mexican herders. Lack of capital continued to inhibit the growth of industries.

In the 1850s the Texas range sheep and goat industry needed a promoter. It found one in George Wilkins Kendall.

3
George Wilkins Kendall

No one did more to promote the Texas range sheep and goat industry in its early days than George Wilkins Kendall. An erect man with piercing eyes that peered out above a round nose, Kendall was a talented journalist, a chronicler of the 1841 Texan–Santa Fe expedition, a Mexican War correspondent, an editor of the New Orleans *Picayune*, an author of books, and one of the most progressive and enthusiastic sheep ranchers Texas has known.

After arriving with sheep in the rugged, cedar-covered terrain of the Texas Hill Country, Kendall spent large, but unrecorded, sums of money from his private accounts in writing essays and articles to northern newspapers describing the benefits of Texas. One famous note, signed with the pen name Acorn and published in the Boston *Post* in 1858, brought a veritable avalanche of letters in reply. Most writers wanted to know about the Texas climate and soil and the best locations for raising sheep and their livestock. During the four months immediately following publication of the Acorn letter, replies to the inquiries cost Kendall over one hundred dollars in cash plus many hours of time.

Born in New Hampshire in 1809, as a youth Kendall was a printer's apprentice. His formal education was brief and haphazard, but he read a lot, especially in geography, his favorite subject. When seven years old, Kendall went to live with his grandfather near Amherst, where for ten years he lived a normal New England farm boy's life, working in the fields, tramping through the woods, and fishing in the rivers and streams.[1] No doubt, too, he became familiar with sheep and wool production.

[1] Fayette Copeland, *Kendall of the Picayune*, pp. 6–9.

Early in 1824 George Kendall went to work in the print shop of his cousin's newspaper, the Amherst *Herald*. Not long afterward, he determined to make printing and newspaper work his career. When the *Herald* folded a year later, he began several years of wandering, catching print jobs here and there. In 1835 he was in New Orleans, and two years later with Francis A. Lumsden, whom he had met while working in Washington for the *National Intelligencer*, he launched the New Orleans *Picayune*. The paper was an immediate success. It became one of the most important newspapers in the South, if not in America, and while Kendall directed it it was a strong promoter of Texas agriculture and the western sheep and wool industry.

Impatient and always restless, Kendall did not remain long in New Orleans. Although retaining part ownership of the *Picayune*, he traveled extensively. Each summer for several years he journeyed northward to purchase paper, type, ink, and other supplies, and in 1841 he accompanied the Texan–Santa Fe expedition to New Mexico. The trip introduced him to his beloved Texas Hill Country and to Mexico, where for a time he huddled behind prison walls—the result of the ill-fated attempt by Texans to place New Mexico under the political jurisdiction of the Republic of Texas. During the War with Mexico, 1846–48, he accompanied General Winfield Scott's forces through the campaign from Veracruz to Mexico City. Afterward he went to Europe, where in France he met, courted, and in 1849 married eighteen-year-old Adelene de Valcourt, the daughter of a prominent French family related to Napoleon Bonaparte.

In 1852 Kendall launched into the Texas sheep business. He began operations with three partners: A. M. Holbrook, a colleague from the *Picayune*, Captain Forbes Britain, whom he had known in Mexico as a war correspondent, and W. A. Weed, a Vermonter he had met in Mexico at the time of the Texan–Santa Fe expedition. Kendall's first trip to San Antonio, in 1841, and his experience with the Texan–Santa Fe expedition apparently convinced him that a profit could be made by grazing sheep in Texas. "I thought," he wrote in the 1858 *Texas Almanac*, "that if in Vermont . . . money could be made by raising sheep, where land was worth from $20 to $60 per acre, and where the farmers were compelled to toil hard six months in the year to keep their flocks alive the other six, that it would be far more remunerative in Texas, where lands could be purchased at from 50 cents to $1 per acre, and where the animals could pick their own living almost the year around."[2] In fact, to herd sheep, he had acquired as early as 1841 a large ranch on the lower Nueces River near San Patricio and during the War with Mexico a smaller one on the lower Brazos.

For two years, however, Kendall did little about his sheep enterprise. He was not satisfied with the Nueces River country as a home, he could not find capable herders, and he spent many months in Paris with his wife. Nevertheless, Kendall and his partners purchased a flock of six hundred *chaurro* ewes, added to it twenty-four purebred Spanish Merinos they had purchased in Vermont, and pastured the animals on the Nueces River property. While Kendall was in France, Weed searched for new land in Texas.

In September, 1852, upon the urgent request of Weed, who was in Corpus Christi, Kendall departed France. Enroute home he stopped in Scotland, where on the banks of the Tweed River in the Cheviot Hills, he hired a young, resourceful herder named Joe Tait, who was to help lay the foundations for Kendall's success in the Texas sheep business. Kendall also bought four collies and two fine Leicester rams. Although the sheep failed as an investment, when they did not adapt to Texas conditions, Tait and the dogs contributed greatly to Kendall's enterprise.

Back in Texas, Kendall, leaving Tait at the Nueces, toured the San Antonio region with Weed, searching for a suitable location. They rode into hills southwest of the city and examined the area near Seguin some thirty-five miles to the east. Then, late in November, in Comal County four miles above New Braunfels, which was a small but thriving German community established in 1845 along the upper Guadalupe River, they found a favorable site. It was, Kendall wrote to his wife in Paris, "a pleasant and verdant valley, surrounded on all sides by rough, rocky and rugged mountains," located in the Texas Hill Country. The tract, he continued, was "about six times as large as all Paris, [and] admirable adapted for raising both sheep and horses."[3] It became known as the Waco Springs Ranch.

Problems followed. Before he could transfer his flock to the new site, an unprecedented blizzard swept the range in early 1853, and many of his animals froze to death in the sleet and snow. Then, later in

[2] *Texas Almanac*, 1858, p. 134.

[3] Copeland, *Kendall of the Picayune*, p. 260.

the year, after the sheep had been moved, the flock did badly, many animals dying from a species of liver rot. By this time, however, Kendall had purchased a small flock from a General Pitts, a sheepman who resided about ten miles away from the New Braunfels site. Fortunately for Kendall, the liver rot disease attacked only animals that had come from the Nueces region, where they had been pastured on rich mesquite grass and on flat, moist soil. The sheep purchased from Pitts, pastured on the same dry, comparatively high country with sandy soil as Kendall's new land, almost entirely escaped. Although several of the Merino rams brought from Vermont died among the rest and the general appearance of the flock was disappointing, Kendall was not discouraged. "So long as I saw," he wrote in the *Texas Almanac*, "that the sheep bought from Gen. Pitts continued healthy, there was hope; they were 'native here, and to the manor born,' and there might be nothing in the soil, climate or pasturage to generate so fell a disease as the liver rot."[4]

There was another problem: Kendall could not purchase the Waco Springs property. When he left again for Europe in early 1853, he instructed Weed to locate the owner of the land in San Antonio. But since the owner, an immigrant from Ireland, had returned to Europe, Weed's task proved impossible. After Weed by letter informed him of the complication, Kendall went to Dublin, Ireland, but he also failed to locate the owner. An agent he hired eventually completed the purchase. Finally, in 1854 a prairie fire swept his land. Although he lost no sheep, he determined to move most of his flock westward to frontier land his partner A. M. Holbrook had acquired that year.[5]

Consequenty, in the summer of 1854, Kendall's workmen, including Joe Tait, a Dutchman, a Frenchman, an Englishman, and several Mexican-American herders, began moving the flock. The site, Post Oak Spring, was a dry, hilly country about six miles east of present Boerne. After locating the sheep, the men built some pens and sheds and set up a small but comfortable camp for the herders. With the sheep grazing at sites thirty miles apart, Tait and his assistants remained busy. They could not always give the animals the close attention normally required.

[4] *Texas Almanac*, 1858, p. 134.

[5] Survey 243 (Juan Ortiz Survey), Land Records, Kendall County Abstract Co., Boerne, Texas.

Nevertheless, the animals fared well. The liver rot seemed to have run its course, and the general appearance of the flocks materially improved. The lambs, the product of crossbreeding *chaurro* ewes and Merino rams, gave signs of upgrading, not only in form and apparent vigor, but particularly in the quality and quantity of the wool. "Here I might state," wrote Kendall, "that a Mexican ewe, shearing one pound of course wool, if bred to a Merino buck of pure and approved good blood, will produce a lamb, which, when one year old, will shear at least 3 pounds of much finer wool."[6]

But problems continued to plague Kendall's efforts. In the winter of 1854–55 another fire, blown before a strong norther, swept over the dry grass. The herder in charge at the time carelessly allowed the sheep to get caught in the blaze. Although it killed no animals outright, the fire so badly burned the sheep about the feet and legs that between 400 and 500 soon died. Early in the spring the charred earth caused further difficulty. Before new grass had started, the ewes in late February began dropping lambs. They could barely pick enough of the old grass to sustain themselves. In mid-March, a cold rain or sleet struck the weakened animals. Many lambs born during the sleet never gained their legs, while the ewes gave but little milk, for lack of new grass. Afterward Kendall determined thenceforth to set his lambing season for the first of April.

Late in the year another problem hit. It proved to be a peculiar species of disease, which Kendall blamed on acorns. There had been a heavy mast of acorns that season, and the lambs, many of which were stunted and puny in early summer for want of milk, devoured the nuts with avidity and probably had not enough strength to digest them. The disease, or whatever it may have been, did not appear again.

The subsequent winter of 1855–56 marked the beginning of Kendall's success. The weather that winter was severe: a succession of cold, wet northers followed each other from January until late March with hardly a day's intermission. Because the grass was poor and Kendall had no fodder or cured hay stacked nearby, the sheep suffered, and many died. But the lambs, except for a few, did not begin dropping until April 1, when warm weather had returned. Losses proved fewer than they might have, had the lambing season developed earlier. "My

[6] *Texas Almanac*, 1858, pp. 134–35.

success since April, 1856," Kendall wrote, "has been most flattering—has been unprecedented."[7]

Kendall prospered indeed. In 1856 he counted some 1,850 sheep and lambs. Two years later, although he had sold many bucks and wethers, he owned more than 3,500 animals, in effect doubling his flock. Over the same period the amount of sheared wool more than tripled, from 2,800 pounds in May, 1856, to 9,000 pounds two years later. In addition, by continuing to breed the ewes to Merino rams, he materially improved the quality of the wool. "I doubt," he enthusiastically reported, "whether a greater degree of good fortune ever attended the efforts of any one engaged in the business."[8]

As he prospered, Kendall spoke glowingly of the Texas range sheep industry. He wrote of the land near Post Oak Spring, believing that the high, dry country in the hilly region of Comal and Blanco counties was ideal for sheep husbandry. The winters were unusually mild, water in the creeks and other tributaries of the Guadalupe River was clear and clean, and the short grass was excellent. He noted the absence of disease: scab and foot rot, he believed, could not generate or spread in the region. Liver rot, one of the worst communicable problems, showed no signs of gaining a foothold after the first outbreak among his sheep from the Nueces River—"we might as well look for a visitation of yellow fever," he said.[9]

Most enthusiastically, Kendall talked about his Merino rams. "A well coated Merino," he reported, "cares nothing for snow, or the coldest wind that blows. Nor does rain or sleet affect this breed." In the summer they withstood the extreme Texas heat, and rain or shine, hot or cold, Kendall penned them in the open air. He heartily believed in the prize rams. When he first entered the sheep business in a limited way on the Nueces, Kendall had secured eighteen full-blooded bucks to upgrade his Spanish *chaurros*. Later, in 1853, when in France on one of his last vacations there, he had purchased at an annual spring sheep auction several fine Rambouillet rams, the French strain of Merino, for shipment to Texas. Every four years afterward he bought additional young Merinos. "The Merino brought to the flock in the proper

[7] Ibid., p. 135.
[8] *Texas Almanac*, 1859, p. 126.
[9] Ibid., p. 127.

time," he claimed, was "the hardiest, the healthiest, the most gentle, the easiest managed, . . . the longest lived . . ." of any fine ram.[10]

As he developed his sheep husbandry by trial and error, Kendall in 1856 brought part of his family to Texas. In addition to his wife, Adeline, there were four children: Georgina, William, Caroline, and Henry. Caroline, a deaf-mute, attended a special school in Paris, and Henry, less than a year old at the time and apparently too young to travel, remained in France. Georgina and William with their parents settled easily into the small but comfortable house Kendall had built on the Waco Springs property. Later Henry joined the family. Waco Springs remained the family home until 1861, when the Kendalls moved permanently to Post Oak Spring.

Kendall's property at Post Oak Spring, thirty miles west of the farm, proved vulnerable to Comanche raids. As a result, his herders maintained a system of armed sentries every night. In 1856 Kendall wrote in the *Picayune* that contrary to the poetic description of shepherds bearing "crooks on their shoulders and perhaps lutes under their arms, [his] gentle folk who tended the flocks" stood ready to stand off a full-fledged Comanche raid.[11] For each flock of eight hundred sheep he provided a herder with a double-barreled gun, a Bowie knife, and a Colt six-shooter. His herders resembled border ruffians. He also encouraged the men to wear as ferocious-looking beards and mustaches as possible. Despite the Indian menace, Kendall maintained most of his flock at Post Oak Spring. Near the campsite he built a small gristmill to complement the sheds and sheep pens. He traveled from the Waco Springs farm to the Post Oak Spring ranch twice monthly to inspect his sheep and grind grain for food.

The journeys were always pleasant ones, and he wrote about them in his dispatches to the *Picayune* and in his letters to friends. His accounts of Texas over the next few years, boasting of its wonderful opportunities and its healthful climate, reflected his own good feeling for the state. Perhaps these years were, as Fayette Copeland suggests, "the most contented years of his life."[12]

By 1858 Kendall's writings about the state had inspired queries.

[10] Ibid., pp. 127–29. The quote is cited by Thos. Decrow, p. 129.

[11] Quoted in Edward N. Wentworth, "The Golden Fleece in Texas," *Sheep and Goat Raiser* 24 (December, 1943): 20.

[12] Copeland, *Kendall of the Picayune*, p. 284. See also Harry J. Brown, ed., *Letters from a Texas Sheep Ranch*, pp. 1–24.

The famous Acorn letter remains the best-known example, but there were others that brought many questioning responses. To facilitate replies, the *Picayune* under Kendall's direction printed circulars describing the Hill Country of West Texas and sent them to its absentee editor. Afterward, upon receiving a letter asking about Texas, Kendall would return one of the new publications, in which he boasted of the easy profit in Texas sheep. He also began in 1858 submitting an annual article on sheep to the *Texas Almanac*, a publication started in 1857 by the influencial *Galveston News*. Kendall published eight articles in the *Texas Almanac*, missing only those years during the Civil War when the publication did not appear. In his pieces, he spoke of his own successful operations, described the work of other enterprising sheepmen, promoted the raising of sheep for wool as well as for mutton, and discussed various difficulties associated with sheep husbandry in the Texas Hill Country. He spared little, noting with equal forthrightness the failures and successes of his business. But he always advertised the good fortune that could be gained by anyone who worked diligently and carefully with Texas sheep. At the time of his sudden death in 1867, most people knew him as a knowledgeable authority on the range sheep industry.

Meanwhile, in 1859 his friends had tried unsuccessfully to run Kendall for Texas governor. William Richardson of the *Galveston News*, with the enthusiastic support of the *San Antonio Herald* and Kendall's neighbors in New Braunfels and Boerne, launched a short-lived campaign. In advancing his name, the *News* indicated that Kendall, one of the best-known citizens of Texas, had "raised a monument to himself through the introduction of fine-fleeced sheep to Texas that would outlive all politicians."[13] But in 1859 he was not interested in learning a new profession or in entering politics. After a meeting with Richardson in May in San Antonio, the brief campaign ended. "I have no taste for the calling of a politician," Kendall said, "have never been in the business, and am too old to learn a new trade." The politicians had not had enough, however. The brief campaign and Kendall's positive accounts of Texas had spread his fame across the state, writes Copeland, as "one of its most romantic and successful figures."[14] On January 10, 1862, by order of the state legislature, Kendall County was

[13] Wentworth, "The Golden Fleece in Texas," p. 20.

[14] *San Antonio Herald*, June 18, 1859; Copeland, *Kendall of the Picayune*, p. 290.

carved from the eastern portions of Blanco County and named for him. Boerne, by vote of Kendall County's citizens, became the seat of government.

At the same time the Civil War and other problems almost wrecked Kendall's sheep ranching. First, the war dried up his wool market, the chief source of sheep-ranching income. Up to 1860 Kendall sold his wool clip in Atlanta, but fighting along the Mississippi closed the Georgia market. In 1861 he loaded 18,000 pounds of wool onto Mexican *carretas* and despite the presence of Federal troops delivered it in New Orleans. The next year he was unable to sell his clip, but in 1863 he sold the previous year's crop to an agent in San Antonio, who freighted it to Eagle Pass and marketed it in Mexico. The 1863 clip remained in storage for months before he sold it in Shreveport, Louisiana. Most of the 1863 clip rotted, and he was not able to sell the 1864 clip until he had shorn his animals in 1865. Similarly, he had problems disposing of rams. Just before the war, a big demand had existed for his stud rams. Kendall sold them all over the state for as much as $250 each. Despite petty red-tape annoyances in meeting regulations of the Confederate and Mexican governments, he sold profitably in 1862 more than three hundred grade rams in Mexico. Afterwards, the market declined.

Mounting indebtedness followed. As he ran short of specie, Kendall found it difficult to retain herders. Although he kept a few slaves, most of the herders and other ranch employees were hired Mexicans or Mexican-Americans (Joe Tait had left in 1859 to start his own operation on the Fuente Frio). Accustomed to receiving their wages in gold or silver, the herders saw the Confederate bank notes as unacceptable substitutes. Several times Kendall rode to San Antonio to find new employees, but they often proved unreliable, and flock losses became disastrous as irresponsible herders abandoned their sheep. During some years Kendall sent his sons to tend the sheep.

Kendall's sons were youngsters, however, only nine and eleven years old when the war started, and the Post Oak Spring ranch was located on the frontier. As the war drew away most of the adult population, Indians became ever bolder in their raids. They ambushed herders, slaughtering them sometimes under a heavy attack with bows and arrows. They ran off sheep, causing some operators, but not Kendall, to lose their entire flocks. The inability of the state government to control the Indians caused Kendall to organize his neighbors into bands similar to Texas Rangers. All proved to little avail, however, and a period of terrorism swept across the Texas frontier. At one time Kendall found it necessary to barricade his family in the ranch house at Post Oak Spring.

Even while the Indians took a heavy toll, the unpredictable weather played havoc with the flocks. For Kendall, 1864 proved a disastrously difficult year. It opened with a blizzard that on New Year's Day drove temperatures well below freezing. Kendall lost thirty-five lambs in the storm. January continued with a norther and a chilling rain that intensified his burdens. February closed with heavy snows, whirling drifts, and freezing cold. The flockmaster found both sheep and herders frozen to death, some of the men located only after a three-day search. In March a drought began, lasting until the end of the year. Creeks disappeared, and grass and feed crops withered. The appearance of great herds of wild Texas longhorns complicated the difficulties. Looking for water, the cattle, neglected during the war and increasing in numbers into the millions of head, grazed in the valleys, destroying planted fields, tearing down fences, and trampling the last sources of water into bogs, which soon became half-dry mud holes and then caked-over wastes. Kendall battled the cattle, killing many of the beasts for food.

Finally in the fall of 1864 there was a serious epidemic of scab, a disease that had first appeared in the winter of 1860–61. It was a cutaneous disorder carried by the acarus, or itch insect. Very contagious, scab spread from one animal to another by contact or by a healthy sheep's touching a tree or post against which an infected animal had rubbed. As the disease advanced, the victim rubbed and bit itself violently, breaking the pustules and hence further spreading the problem. Scabs formed over red inflamed sores, and between the sloughing action of the sores and the teeth of the sheep, the fleece eventually hung in tatters about the body. If allowed to run its course, scab resulted in death. Contemporary treatment called for direct application of grease or for the use of tobacco extractions.

In 1861 his scab problems proved minimal and Kendall did little. But in 1864, when scab hit his flock hard, he attacked the problem with vigor. From early October to late January, 1865, he fought the disease, treating only infected animals first before deciding to treat his entire flock. Unable to purchase either lye or lime, he made his own by burning rock. He hired extra help, collected material, and planned his fight. Because his flock numbered five thousand, Kendall, rather than spread the medicine on the animal, planned to dip each sheep in one of several kettles and barrels borrowed from neighbors, containing a liquefied mixture of the remedy. Into the vessels he dipped the sheep—a tedious process. At first he mixed too much lye, causing the hands of two of his men to bleed. When that difficulty was solved, a winter norther blew in, with freezing temperatures and a cold rain. Work slowed but did not stop until the twenty-first of January, when the entire flock had been dipped.

The misadventures led to more careful planning. In the spring, when shearing was over, Kendall prepared to dip his animals again. This time, however, he built a large vat or trough into which his hired hands drove the sheep. The vat stood nine feet long, three feet wide, and four feet deep, sunk into the ground three feet. From the bottom of the vat about three feet from one end, some ten or twelve steps or stairs led out upon a platform eight by thirty feet, sloping so that after the sheep entered the solution-filled vat they climbed onto the platform and drippings from them ran directly back into the trough. Kendall constructed his trough and platform of rock and stout plank. On the end of the vat opposite the platform, he erected a pen to hold one hundred head. From the pen his men let the animals gently down into the trough on their backs. After they had risen to the surface, the animals floundered and splashed before turning to find their way onto the platform—and the performance was over. In this way Kendall kept two or three grown sheep splashing at a time. From the upper end of the platform, through a small gate, he led the animals into a receiving pen. The scab pestilence eventually declined, but by his use of the big vat Kendall had led the way in initiating large-scale dipping operations.[15] After the Civil War western sheepmen built large dipping vats modeled on Kendall's plan, and commercial firms began producing scab medicine for use in the vats.

When the Civil War ended in 1865, wool prices improved. In the fall wool sold for twenty-four cents a pound, and outlets for its sale be-

[15] *Texas Almanac*, 1867, pp. 217–18. See also G. R. Bode, "The Life and Times of George Wilkins Kendall" (M.A. thesis, University of Texas, Austin, 1930).

came widely available for the first time in years. As market conditions improved, Kendall prospered and took his family on an extensive trip. They visited New Orleans, the east coast, and France. By now Kendall was well known among wool growers. His prestige had led before the Civil War to a close relationship between himself and Henry S. Randall of New York, the author of a lengthy biography of Thomas Jefferson, a nationally recognized sheep authority, and a long-term president of the National Wool Growers Association.[16] In New York Kendall and Randall reestablished their old relationship.

After the trip Kendall returned to Post Oak Spring, where following a short illness he died on October 21, 1867. For some seventeen years the famous newspaperman had engaged in sheep raising as a serious business. During the period he had enjoyed much success and suffered many reverses. He had imported purebred rams from Europe and developed range sheep that produced a profitable wool clip. His grade rams had been distributed throughout Texas and had helped to improve the breed of many flocks. He had combated disease and vermin with the dipping vat. He had experienced all of the normal troubles of sheep raising plus others inspired by the Civil War and cattle competition. More than any other person, he had shifted the industry's emphasis from mutton to wool production. When he died, George Wilkins Kendall left no monetary fortune, but his death marked the passing of one of the state's greatest sheepmen at a time when postwar wool prices were unleashing a boom in the sheep industry of the Rio Grande Plain.

[16] Brown, *Letters from a Texas Sheep Ranch*, pp. 1–24.

4

The Rio Grande Plain

THE Rio Grande Plain, or Wild Horse Desert as it was called by most people in the mid-nineteenth century, dominated the Texas range sheep industry for two decades after the Civil War. As George Wilkins Kendall struggled to raise sheep and produce wool in the Hill Country during the 1850s, sheepmen were trickling into the Rio Grande Plain. In the early 1860s the trickle became a stream and after the Civil War a flood. The boom did not last, and in the late 1880s the sheep and goat industry in the Rio Grande Plain declined, although not before it had played a role in Texas agriculture that at times overshadowed the colorful South Texas cattle industry.

The Rio Grande Plain in extreme South Texas consists of approximately 22 million acres in all or parts of some thirty-five counties bounded on the north by the Balcones Escarpment and the San Antonio River, on the east by the Gulf of Mexico, and on the south and west by the Rio Grande. Much of the land is drained by the Nueces River and forms a gently undulating plain sloping to the southeast about 200 to 700 feet above sea level. An almost level belt of land 180 miles long and 30 to 60 miles wide parallels the Gulf coast. A narrow range of hills up to 1,000 feet high parallels the upper Rio Grande.

The climate is mild. Average annual temperatures vary from sixty-eight degrees (Fahrenheit) in the north, around San Antonio, to seventy-four degrees in the extreme lower Rio Grande Valley. Summers are hot, daytime temperatures commonly topping ninety degrees. Winters are mild, although cold fronts periodically bring freezing temperatures. Annual rainfall averages about thirty-three inches in the eastern part of the region but gradually decreases westward to approximately twenty inches on the upper Rio Grande.

Vegetation varied in the nineteenth century. Post oak and hickory comprised woodlands, some quite extensive, from San Antonio south and east to Goliad. Along the Gulf coast tall-grass country provided useful cover and some feed for large numbers of deer, the most common native grazing animal in the Rio Grande Plain. West and north of the flat coastal plains short grasses predominated. Islands of cactus and shrubs also grew there, but one could find stands of live oak and mesquite, quite noticeable at the southern edge of Wild Horse Desert in northern Willacy County and in the Nueces sands of Kenedy, Brooks, and Jim Hogg counties. Mesquite, ash, elm, and other trees grew along many of the large creeks that flow toward the Gulf. Grasses and scrub brush, however, dominated the area. In the extreme lower Rio Grande, palms, mesquite, anaqua, ebony, and other trees grew abundantly. Short brush and semidesert shrubs covered the rolling hills paralleling the river from about Rio Grande City upstream.

Perhaps fifty thousand sheep and goats in 1850 roamed over the vast rolling Rio Grande Plain. Disruptive activities of Mexican, Texan, and Indian raiders and other pressures in 1800 and afterward prevented the growth of larger numbers. A few sizeable flocks grazed near Corpus Christi, below San Antonio, and amid some of the old settlements along the Rio Grande, but elsewhere the animals were more scattered, belonging to Mexican herders working small, isolated flocks. In fact, General Zachary Taylor, leading his United States Army southward from Corpus Christi toward the Rio Grande in 1846, when the War with Mexico was about to start, found, except for wild horses, few animals of any kind.

Restoration of peace at the close of the war in 1848 brought scores of new stockmen into South Texas. The new arrivals plus the return of many people who had fled extreme South Texas during the war launched the Rio Grande Plain into a half-century of large-scale sheep and goat production. Most sheep entering South Texas in the 1850s and afterward came from the plains of Tamaulipas, Coahuila, and Nuevo León in Mexico. Ranches there provided initial flocks to Texas sheepmen and continued to provide replacement stock after sheep production was well established. Additional animals were brought in by settlers from the east and northeast. They were responsible for a considerable influx of purebred stock, especially Merino, Southdown, and Delaine, the American Merino strain. Some purebred sheep arrived by boat at Brownsville, Indianola, or Corpus Christi and then were driven or hauled to the South Texas ranches and sheep camps.

In 1850 the United States census reported more than 100,000

sheep in Texas—less than one percent of the country's total, but half of that number grazed below San Antonio. Those in the extreme lower Rio Grande were primarily the descendants of the Spanish *chaurro*, the long-legged, large-framed meat breed that produced only a pound or two of coarse wool. Although this mutton breed dominated the Rio Grande Plain until the 1880s, raising sheep for wool was an attractive proposition from the first. Prices varied from year to year, but the demand remained consistent. Moreover, South Texas wool, the first crop sheared each spring, usually commanded a premium, and selling was easy. Consequently, even in the early days of the Rio Grande Plain sheep industry, economic circumstances dictated the building of ever larger flocks to produce more wool. In addition, producing wool entailed only small financial risks, for it was believed that if sheep died before they were six months old, they had cost the raiser little. If they died afterward, their wool had more than paid for what they had consumed in feed.

Plenty of markets existed. Each year Americans drove a few sheep across the river to Mexico, and Mexican buyers brought up wagons from Chihuahua and elsewhere to carry away the wool. Inhabitants of the lower Rio Grande used wool for ponchos and coarse rugs. Small amounts of wool left Brownsville by ship for overseas. Farther north, wool sold in San Antonio, Brazos de Santiago, Saluria (once a port on the northern tip of Matagorda Bay), and Corpus Christi, which would export for a time more than one-third of all sheep and wool annually leaving the United States. In addition, there had developed in the upper Guadalupe River Valley, north of San Antonio, heavy local demand for sheep for breeding stock.

Such opportunities attracted many people. One of the first to introduce fine Merino sheep into the section was Israel T. Bigelow, a judge in Brownsville, who in the winter of 1848–49 brought to the mouth of the Rio Grande some eight or nine rams and crossed them to *chaurros*. Bigelow kept his sheep near Brownsville two years and then removed the flock, numbering about 450 head, to near Tio Cano. The sheep, for unknown reasons, declined after the move, forcing Bigelow in 1854 to sell the remaining animals.[1]

Late in 1849 Major W. W. Chapman, a quartermaster in the United States Army stationed at Fort Brown, imported some sixty

[1] *Corpus Christi Weekly Gazette*, January 10, 1874.

head of Merino rams and ewes. He purchased the sheep in Washington County, Pennsylvania, long a favorite buying place for Texas sheepmen, from Thomas Ewing. In company with his friend and business partner Charles Stillman, Ewing brought several small lots of Merino rams to the Rio Grande Valley by boat in 1850, 1851, and 1852. The three men established a ranch at Jarrita, near Arroyo Colorado, and placed the animals under the care of Jorge Cavassos, a skilled Mexican herder. In 1853 their flock numbered over ten thousand head. The following year, when they lost three-fourths of their sheep from heavy rains and severe winter weather, they abandoned sheep raising in the valley. Chapman, however, transferred in 1853 to Corpus Christi, imported some fine Southdown and Bakewell (Leicester) sheep, and inaugurated the distinguished Chapman Ranch in Nueces County.

In 1850 Captain Mifflin Kenedy and Richard King, partners in a steamboat operation on the Rio Grande before settling into ranching along the Texas Gulf coast south of Corpus Christi, bought 50 head of Merino rams in Pennsylvania. They transported the sheep by boat down the Ohio and Mississippi rivers. The *South America*, the steamer carrying the animals, caught fire near New Madrid, Missouri, and burned, destroying everything on board. Kenedy and King then bought fine wool sheep locally to breed with the common Spanish ewes. In his two-volume history of the enormous King Ranch, Tom Lea quotes from an 1854–55 ranch account book: "paid E. D. Smith on 376 sheep . . . $282.00; . . . cash paid for sheep and goats bought through Lewis Mallett, $145.00, charges and expenses on same $13.00; H. A. Caldwell for 130 goats $97.50." Within twenty years, the King Ranch sheep stock numbered 30,000 head or more, all Merinos or crossbred Merinos. They yielded an average of four pounds of wool per head. King placed his sheep under the care of Captain J. S. Greer, who, according to the *Corpus Christi Weekly Gazette*, took "great pain and pride in their successful culture."[2] From King's main sheep ranch, called Borregas, Greer distributed the animals to best suit the grazing grounds. Greer, unlike many of his contemporaries, ran sheep and goats with cattle and horses.

Perhaps the biggest importer of fine Merinos was John McClane. His father raised pureblood Merinos in Pennsylvania, selling hundreds

[2]Tom Lea, *The King Ranch*, I, 122, 304; *Corpus Christi Weekly Gazette*, February 7, 1874.

of sheep to South Texas ranchers. Young McClane, accompanying a band of sheep headed for Texas, arrived in Corpus Christi in the early 1850s. There he served as an agent for his father and built up his own flocks in Nueces County. He became quickly acquainted with many of the largest sheepmen in the region. He persuaded Captain Kenedy to purchase seventy-five Merino rams from his father at a cost of $3,500. Only fifty-six of the rams landed at Point Isabel in Cameron County, the others having drowned in a storm in the Gulf of Mexico. Kenedy's flock in 1854 numbered over ten thousand head, but after heavy floods and rains, together with the same severe weather that devastated W. W. Chapman's sheep, most of the flock perished. Two years later Kenedy sold what remained of his flock to McClane, who brought them to his own spread in Nueces County. McClane, who helped to popularize the importation of fine-wooled sheep, added more Merinos and after six years found his sheep valued at $60,000.[3]

For these successful ranchers, the start of sheep and wool production caused no problem. Each plumped down a handful of money, hired a herder or two, and watched his flock increase. Sometimes, however, getting started proved more difficult. In the Rio Grande Plain a common beginning included about one thousand ewes, costing about $1,250, and six to ten rams, costing $100. A flock of this size could be managed by a single *pastor*, or Mexican herder. Another $25 purchased a small herd of goats to furnish the *pastor* meat. Then grazing land was needed, amounting to one thousand acres for each one thousand sheep. The usual cost of leasing land was about three cents per acre yearly. The *pastor* received as wages between $5 and $10 per month. Finally, there were a shelter for the herder and a monthly grub ration to augment the goat meat. The annual cost of store-bought provisions, including three pounds of coffee, thirty pounds of flour, three pounds of sugar, beans, salt pork, and other staples, reached about $100. Shelter was not an expensive item. According to Sandy McNubbin, an early Scot sheepman in the region, lodging was "sometimes a pile of brush, and sometimes a little lumber house about five feet by seven feet was given in which to store provisions."[4] More often a rag tent sufficed.

[3]John Ashton, "The Start of Sheep Breeding in Texas," *Sheep and Goat Raiser* 25 (December, 1944): 36.

[4]*Corpus Christi Caller*, July 19, 1925. One sheep per acre meant an overgrazed range. But in the mid-nineteenth century there was little concern or little knowledge

The *pastores* in the Rio Grande Plain led a serflike existence. In perpetual debt to their masters, they were miserably clad in tattered blankets, worn shoes, patched clothes, and brimless hats. Sometimes armed only with bows and arrows, they were nevertheless brave, self-sufficient, and dedicated to the well-being of the sheep. For months at a time their only companions were a dog or two and the *vaquero* who made the rounds of the herders' camps every few weeks, bringing fresh food supplies, reading material, tobacco, and messages from the owner of the sheep. An experienced Mexican herder had to be bought from another sheepman for around a hundred dollars. Indeed, the *pastor* found it difficult to get a place as a herder without a large indebtedness, for a Mexican sheepherder out of debt, it was unfairly believed, was a poor risk. The new employer willingly paid or became responsible for the debt.[5]

Since he had little money to hire *pastores* or purchase sheep, the average American or European newcomer to the sheep business operated under the *partido* system. Under such a scheme large ranchers or companies, especially those in Corpus Christi, farmed out their sheep on shares. Often the herder would be required to return yearly sixteen or twenty head for each one hundred advanced, and the owner and the herder would similarly share in the annual wool clip. The herder accepted responsibility for the health and safety of the flock and performed all the many chores associated with sheep husbandry. The system assured the rancher that the herder had a strong incentive to take care of his sheep, and it enabled the herder to work toward his dream of one day becoming an independent rancher himself. In fact, in such a manner scores of sheepmen got their start.

Although sheep raising was initially expensive, potential profits from the sheep business were considerable. One authority wrote in 1880 that annual profits from sheep and wool ranged from about 21 percent to 75 percent. A recent account based on early accounts has estimated that an investment in sheep could be doubled annually. Such reports seem overly optimistic, but the stories were typical of nineteenth-century reporting. Val Lehmann in *Forgotten Legions*, a de-

---

about conservation of range land, and one sheep per acre was the generally accepted standard.

[5]Coleman M. Campbell, "Era of Wool and Sheep in Nueces Valley," *Frontier Times* 2 (February, 1934): 196.

tailed study of sheep in South Texas, catalogues more than half a dozen similar accounts of hoped-for profits.[6]

James Bryden actually enjoyed such large profits. With a few friends he established James Bryden and Company, a sheep-raising and mercantile concern in Nueces County. In April, 1856, his outfit located a flock at New Santa Gertrudes. The flock consisted of 694 ewes and 11 rams, costing on the average less than $2 per head. Four years later the company counted 3,778 head, despite having sold 599 rams and wethers and having lost 199 sheep. Bryden's firm also sold four wool clips during the period, estimating its proceeds from the sale at more than $4,000. The sale of the rams and wethers netted over $1,500. For the remaining sheep in 1860 the company received an offer of $5 per head. With these figures as a basis of calculation, the original flock cost about $1,400, and the 1860 flock was valued at nearly $19,000. Add the money from the sale of wool and sheep, and value totaled about $24,500. Subtract from this amount the cost of the original flock, and it leaves a profit of $23,000. Bryden noted that there were expenses of maintaining a *pastor* for the flock, "but this outlay to us is trifling."[7]

As more and more people turned to herding as a livelihood, for almost three decades the number of sheep in the Rio Grande Plain increased. Of course, cattle raising remained important, especially in the northern, eastern, and southeastern edges, where such awesome spreads as the Richard King and Mifflin Kenedy ranches developed. But the great interior of the region was stocked, as Val Lehmann put it, with "forgotten legions of Mexican and Mexican-cross sheep."[8]

Sheep ranches appeared in all the counties of extreme South Texas. George Reynolds at Orange Grove in present Jim Wells County; Thaddeus M. Rhodes, Peter and Albert Champion, and Nestor Maxan in Cameron County; and Dow Hardin in Webb County were among the earlier producers of sheep, goats, and wool. In Webb County one man entered the sheep business in 1855 with 250 ewes. Five years

[6]Clarence Gordon, *Report on Cattle, Sheep and Swine, Supplementary to Enumeration of Livestock on Farms in 1880*, pp. 969–70, 981–83; Hiram Latham, *Trans-Missouri Stock Raising, the Pasture Lands of North America: Winter Grazing*, p. 67; Val W. Lehmann, *Forgotten Legions, Sheep in the Rio Grande Plain of Texas*, p. 39.

[7]James Bryden, "Sheep Raising in Nueces County," *Texas Almanac*, 1861, p. 171.

[8]Lehmann, *Forgotten Legions*, p. 48.

later, having bred his sheep twice yearly, he sold from his flock 3,000 head.[9] As the numbers multiplied, even the big cattle operators grazed sheep with their bovine herds or placed them in separate pastures. In the Rio Grande Plain no significant antagonism existed between cattlemen and sheepmen. Indeed, most ranchers grazed both animals.

Sheep ranches in the Rio Grande Plain varied in size and complexity. Most of them were small, running at most one thousand to two thousand sheep and goats. Owners of small spreads grazed and tended their own flocks, managing with their families the shearing operations and caring for the young animals during the busy lambing season. Occasionally it was necessary to hire one or two men to build pens and a camp or help with the shearing. But profits at the end of the year could reach 75 percent on the original investment.

Some ranches covered tens of thousands of acres. According to one 1880 newspaper report, 300,000 sheep grazed on a 300,000-acre ranch—noted as the largest sheep ranch in the United States—in northern Webb and southern Dimmit counties.[10] Although that report cannot be confirmed by other sources, some sheepmen pastured more than 100,000 animals. William and Robert Adams and Charles Callaghan ran two of the largest operations. The Adams brothers, from a large family of English immigrants, came to Corpus Christi in the 1850s. In 1863, in the midst of the Civil War, when they were about seventeen years old, their father took over on shares a flock of sheep owned by the mercantile firm of Gilpin and Belden in Corpus Christi. Under terms of the agreement, the elder Adams was to herd and care for 3,000 sheep for three years. In exchange, he would receive during the period one-half of the proceeds from the wool clip. At the end of three years he was entitled to one-half of the increase of the flock. William and Robert assisted their father.

Forty miles from Corpus Christi on open range in present Jim Wells County, near Casa Blanca, the Adamses took charge. As agreed, they tended the flock, which they had assumed from one Antonio Maya, an ex-soldier under Santa Anna who had been stationed at the old Mexican fort of Lipantitlan on the Nueces River. They directed the wool clips and bore all responsibilities associated with good herding.

[9] *Texas Almanac*, 1868, p. 113.

[10] *Galveston Daily News*, October 16, 1880.

Although they grazed the sheep over a wide area, they maintained a headquarters camp along Agua Dulce Creek. The Adamses prospered. At the end of three years they possessed their own flock of six hundred sheep and had saved some money from their share of the wool clips, which had sold for between twelve and twenty cents a pound. Early in 1869 they secured 320 acres of land on Tecolote Creek, about fourteen miles north of present Alice. Here they brought in more sheep and, aided by Mexican herders, took over much of the range country in the area. In a few years, by close and intelligent attention to business, the Adams brothers, said one contemporary, "became the owners of eight thousand of the best sheep in the state."[11] Through the 1870s and 1880s they continued to expand their sheep production, but in 1887 with most of the other South Texas ranchers they began to turn away from sheep in favor of cattle raising.

Even larger was the Charles Callaghan operation. In the late 1860s Callaghan, a veteran of the Confederate army, bought 5,000 Mexican *chaurros*, hired several *pastores*, brought in blooded rams to upgrade his ewes, and employed Colonel William R. Jones to manage his property. By the late 1870s the Callaghan ranch, lying astride the Old San Antonio Road near Encinal, held some 100,000 sheep and 6,000 goats on its veritable kingdom of territory. The wool clip in 1881, totaling nearly half a million pounds, represented better than 20 percent of the wool shipped from Encinal, one of the largest marketing points in the state. Callaghan's heirs as late as 1900 continued to raise thousands of sheep, but after 1908 new owners of the ranch turned to cattle operations.

Large operations like Callaghan's—those of 20,000 to 50,000 or more sheep—had a careful hierarchy of organization at work. In the Rio Grande Plain the Spanish system prevailed. At the top, in the absence of the owner, stood the ranch superintendent or foreman. Below the foreman was the *mayordomo*, a man of long experience in the conduct of the practical part of sheep raising. He kept constantly on the range, going the rounds of the different camps, noting the conditions of the sheep, and suggesting changes of ranges. He often held final authority on matters of the sheep and range. Answering to the *mayordomo* were the *caporales*, usually three to a *mayordomo*. The *capo-*

[11] Agnes G. Grimm, *Llano Mestenas, Mustang Plains*, pp. 85–89; W. G. Sutherland, "Adams Bros., Trans-Nueces Pioneers," *Cattleman* 17 (June, 1930): 20.

rales provided monthly reports, rode the range, and saw that their camps and their subordinates were supplied with provisions from the home ranch. They oversaw the movement of about 18,000 animals each and directed the work of three *vaqueros*, each of whom in turn supervised three *pastores*, lowest in the hierarchy. Each *pastor* had charge of from 1,000 to 3,000 head of sheep, which he accompanied by day and camped with by night, moving on foot and assisted by a dog. The *vaquero* exercised close watch over the flocks in his charge, provided a monthly report to the *caporal*, and delivered supplies and news to the *pastor*. On the Callaghan ranch there were fifty herders, eighteen *vaqueros*, and six *caporales*. With smaller flocks the organization differed. Americans streamlined the system to meet their needs, eliminated some rankings, and adopted a few new terms, such as rustler for *vaquero*.

For the *pastor* daily work was largely routine. For him tomorrow brought no difference from yesterday. At break of day the herder took his flock to graze, the dogs attending him. In warm weather he and his dogs slept for about four hours in the middle of the day. At sundown he brought his flock to camp and cooked his supper of coffee, tortillas, and goat meat. Sometimes a little variety interrupted the routine. If the sheep spooked at a wild deer and stampeded, he had to round them up. If he trapped a turkey, quail, or rabbit, he changed his diet of *cabrito*. From this monotonous system he seldom escaped. Many a *pastor*, however, received an annual leave of up to a week. Then, with a horse and saddle, a good hat and probably a pair of boots and twenty dollars spending money, he started out for Mexico, wasting his "substances" in riotous living. He patronized bullfights, fandangos, and monte dealers, drank mescal, and ate chili con carne and tamales. At the end of the third or fourth day his money had disappeared, and the monte dealers had relieved him of his horse, saddle, hat, and boots. He trudged home to his sheep camp, assured of a plentiful supply of food. He knew that his employer would furnish him with shoes and clothes, and he needed neither money nor a horse. Some *pastores*, if given more than a week's reprieve, were known to travel to Mexico City.[12]

[12] *Corpus Christi Caller*, November 30, 1930. See also Edward N. Wentworth, "The Evolution of Sheep Shearing in America," *Sheep and Goat Raiser* 25 (June, 1945): 24–27, 30, 32; and Steven Powers, *The American Merino: For Wool and for Mutton*, p. 225.

In the Rio Grande Plain there were two shearing seasons each year: April to June and August to September. Most owners hired *tasinques* ("shearers") from Mexico, the same men coming in bands to the ranch continuously. The shearers operated in much the same way as modern itinerant wheat harvesters, moving from site to site with the season. Captains organized each band in Mexico. According to the prevailing relationships, each captain provided transportation, maintained order, and made contracts on pay, food, and lodging directly with the rancher or with agents in nearby towns. The captains paid the *tasinques* half of what the owner gave for each animal shorn. Normally prices for shearing a grade sheep ranged from three to four cents per head. An expert *tasinque*, expected to fleece one hundred sheep, seldom made four dollars per day. Some shearers complained that at times they did not get half their earned wages and that the food was poor. On the other hand, notes one historian, the captains claimed that they "traditionally advanced $50 to $400 to a shearer at the beginning of every year, and that once work began, they risked the possibility that the shearer might not work."[13] Captains also provided the shearing equipment.

The shearing season in South Texas as elsewhere in the West represented a time of intense activity, work lasting from daylight until dark. It also represented a few brief weeks in the herder's year when he had the companionship of other sheepmen. It was the "social season" of the sheep industry. The arrival of crews always signaled a major break in the tedium of sheep-ranch life—an excuse to gather for visiting, feasting, drinking, gaming, and probably a little bloody fighting.

Shearing was also sweaty, cumbersome work. In the days before electrically powered shearing equipment, shearing was done by hand in huge barns or in large pens covered with brush or reeds. "At daybreak," notes W. G. Sutherland, "all hands assembled in the pen. Every man had a number of rawhide strings to tie the feet of sheep as he laid them down ready for shearing. Each man as a rule tied ten sheep at a time."[14] Then shearing began. Standing nearby was a boy ready to apply a mixture of charcoal and oil to the wounds of any animal

[13] Arnoldo De Leon, "*Los Tosinques* and the Sheep Shearers Union of North America: A Strike in West Texas, 1934," *West Texas Historical Association Year Book* 55 (1979): 6. See also Stuart Jamieson, *Labor Unionism in American Agriculture*, pp. 225–26.

[14] *Corpus Christi Caller*, November 30, 1930.

accidentally cut. While some men kept tally of the number of sheep tied down by the shearers, others stood at a table with twine to tie up each fleece before it was carefully placed in an eight- to ten-foot sack, which when filled weighed about 360 pounds. In such fashion shearing continued until all animals were trimmed and the sheep crew moved on to another ranch.

Semiannual shearings prevailed only in the Rio Grande Plain. One reason South Texans could afford to shear twice was that Texas short fine wools had an excellent reputation in the wool trade for their felting qualities. Besides, in the hot climate semiannual shearings were beneficial to sheep, especially lambs, promoting their health and general condition. Twice-a-year shearings also offered the owner better opportunity to hold in check and eradicate scab, the disease that wasted many flocks. Moreover, they put money in the rancher's pocket twice a year, a distinct advantage in a region where ranchers frequently asked merchants to advance money on fleeces still on the sheep's backs.

There were other activities: castrating, docking, and marking. In castrating male sheep in the nineteenth century, a worker held the back of the animal against his own chest with all four legs gathered together and elevated. After the end of the scrotum was cut, both testicles were pulled out and the cords severed with a knife, a sharp tug, or with teeth. The latter method, an Old-World practice, was looked upon as a sort of badge of the order, a proof of professionalism. Although there usually was no medication, some salt, charcoal, or oil might be rubbed into the wound. The operation had two purposes: to produce better mutton and to improve the breed by ensuring that the ewes would be bred only to prize rams.

Docking involved removing for sanitary reasons all but about two inches of the tail. Heavy knives, hatchets, chisels, or red-hot docking irons were used. The task was accomplished quickly and cleanly. After some of the loose skin near the base of the tail was forced toward the rump so that it would cover the stump when again in normal position, the tail was severed. Docking was also done for reproductive reasons.

Marking was done for identification. Sheepmen painted their marks on the freshly shorn sheep. Sheep were also earmarked by notching one or both ears with distinctive combinations of cuts and slashes known as "crops," "lance points," or "downfalls." By using a slightly different mark each year, a rancher could tell the age of his sheep when separating them for later market. Owners registered earmarks in the county courthouse. Often brands, likewise registered, were placed on the sheep's jaw or cheek or across the animal's nose.

Market conditions for sheep and wool of the Rio Grande Plain fluctuated but overall tended to improve. Many things affected actual sales, but the Civil War, 1861–65, proved particularly influential. The first effect of war was an increase in the price of wool, and this stimulated expansion of the sheep industry. Because it was difficult to get cotton, many New England textile mills turned to manufacturing wool, supplying it to their regular customers as well as the United States Army, which sought woolen uniforms and blankets. As wool production increased, the economic laws of supply and demand took effect, dropping prices.

The Civil War created other problems, too. Economic conditions in the Confederacy weakened sales there, and when the Union armies captured the Mississippi River, southern markets withered. Moreover, the war, stimulating sheep and wool production above the Ohio River, shut off possible markets in the North. As previously mentioned, George Wilkins Kendall, in response to such developments, applied for permission to export three hundred grade rams to Mexico. He also shipped nearly 19,000 pounds of wool to Eagle Pass, where Mexican buyers paid in specie, which his herders demanded as wages. Later Kendall traded wool in western Louisiana for food and supplies, but for most of the war he could only allow his flocks to increase in number while he stored the wool for sale in more propitious times. Many sheepmen in South Texas quit the sheep industry during the Civil War, ruined by lack of markets and low wool prices.

After the Civil War, when prices in other parts of the United States declined sharply, the Rio Grande Plain sheep business prospered. Heavy local demand for breeding stock was the reason, as thousands of pioneer settlers from Europe and other American states drifted into Texas. Influenced by the success of other sheepmen in the state, they invested in the timid animals. As they bought flocks and leased land, sheepherding in the state spread westward and southwestward. At the end of the war perhaps half a million head of sheep grazed in the Rio Grande Plain, a startling jump from the 50,000 or so of 1850.

As the Rio Grande Plain sheep industry prospered, Corpus Christi grew as a major wool market. From the 1860s to the 1880s wool from all over South Texas, as well as from northern Mexico, found its way by ox wagon or large, two-wheeled Mexican *carretas* ("carts") to the port city. From Corpus Christi wool went to New Orleans, Philadelphia, Boston, or other eastern cities. As a market for wool, hides, and skins in the 1870s, Corpus Christi was a busy port, teeming with side-wheel steamers and schooners. Some wool merchants in the town, notably George F. Evans, J. B. Mitchell, David Hirsch, Uriah Lott, Ed Buckley, and M. Headen and Son, operated on a grand scale, owning spacious warehouses where they stored and sorted wool before shipping it. Some of them had their own ships, which regularly plied the coast between Corpus Christi and eastern seaboard ports. In 1868 the city shipped 1,402,000 pounds of wool and 85,713 pounds of goat skins to New Orleans alone. The numbers increased rapidly thereafter, and in the 1870s warehousemen in Corpus Christi bought and shipped annually from 7 million to 12 million pounds of wool. Years later W. S. Sutherland, a Nueces County sheepherder who had immigrated to Texas from Scotland, noted that Corpus Christi in 1880 "was the greatest wool market in the world."[15]

One of the largest wool commission men in Corpus Christi was Uriah Lott. Born in New York in 1842, he found his way to Texas in the late 1860s, taking a job in one of the wool warehouses in Corpus Christi. He established his own commission business (Uriah Lott and Company) in 1871 and that year chartered three sailing vessels to transport wool to the East Coast. He built on Water Street a warehouse, eighty-five feet long, twenty-five feet wide, and thirty-five feet high, to complement the one he was already operating. A Corpus Christi paper in 1871 reported that "the amount of produce purchased by this firm is exceedingly large: on the schooner *Rosewell* alone they shipped recently 75,000 bags, 5,000 hides and 70,000 pounds of goat skins."[16] Later, with the help of Mifflin Kenedy and Richard King, Lott built the narrow-gauge Corpus Christi and Rio Grande Railroad, called Lott's Folly, from Corpus Christi to Laredo.

Hundreds of oxcarts carried freight between the Corpus Christi warehouses and points in the Rio Grande Plain. Freighters from Mexico brought wool, hides, lead, coffee, and silver ore. They took back

[15] Ibid.

[16] Quoted in Ashton, "Start of Sheep Breeding," p. 42.

with them dry goods, groceries, hardware, and lumber. In May, 1871, one train of fifty carts was enroute to Corpus Christi from the Rio Grande. There were many others that year, too. The street in front of the wool warehouse firm of Mitchell and Evans was almost constantly blocked with carts, either delivering wool or loading merchandise for the return trip. The busy activity kept up through the fall. In November a train of twenty-two carts brought to the firm of Uriah Lott and Company nearly five thousand bags of wool and three hundred of hides.[17]

By 1880 a frenzied sheep boom had hit the Rio Grande Plain. Prices for wool and sheep had reached exciting levels. Texas wool sold for better than twenty-five cents a pound and range sheep marketed for nearly five dollars a head. Grassland was relatively free, and all a speculator had to do was organize a flock and strike out for grazing ground. Almost at once hundreds of prospective sheepmen gathered flocks, hired herders, and spread out through the empty lands of South Texas. Scotsmen, Basques, Englishmen, and others competed with Americans and Mexicans for grasslands and water rights. Promoters of all kinds entered the field. Wealthy eastern individuals who had never been to Texas entrusted their sheep to western adventurers. Although figures vary, according to A. W. Spaight, Texas commissioner of insurance, statistics, and history, sheep and goats in 1882 totaled 2,461,088 animals, numbering more than three times the figure for cattle (see table 1).[18] Webb, Encinal, Duval, and Starr counties were the leading sheep producers, but Maverick, Kinney, and Uvalde counties were not far behind.

Most of the wool went to Boston. Most of the sheep were kept on the range to increase the size of the flock and consequently the size of the wool clip. But as the Rio Grande Plain sheep industry overextended itself, many flocks were trailed to new homes on the Edwards Plateau west and north of San Antonio, or beyond to the Big Bend. In the 1880s South Texas sheep also helped stock the sheep ranges of the

[17] Ibid., pp. 39–42.

[18] Newspaper clipping, Morgue Files, Texas Sheep and Goat Raisers Association, Records, Southwest Collection, Texas Tech University, Lubbock; Lehmann, *Forgotten Legions*, pp. 152–54. For different figures see Samuel Lee Evans, "Texas Agriculture, 1880–1930" (Ph.D. diss., University of Texas, Austin, 1960), pp. 266–67; and U.S. Bureau of the Census, *Tenth Census. Productions of Agriculture, 1880*, III, 170–72.

TABLE 1
Cattle and Sheep and Goat
Populations in 1882

County	Cattle	Sheep and Goats
Aransas	16,204	1,819
Atascosa	13,693	27,888
Bee	39,613	43,346
Bexar	14,062	23,762
Cameron	23,210	32,709
DeWitt	46,597	71,492
Dimmit	11,747	144,070
Duval	11,147	340,559
Encinal*	3,604	228,436
Frio	32,759	26,466
Goliad	58,054	12,012
Hidalgo	16,925	28,120
Karnes	37,115	22,734
Kinney	11,050	107,441
La Salle	16,185	167,910
Live Oak	15,980	48,566
Maverick	3,527	139,448
McMullen	15,616	89,003
Medina	32,699	34,349
Nueces	84,989	182,518
Refugio	84,884	6,548
San Patricio	62,167	3,103
Starr	21,389	156,084
Uvalde	19,802	126,579
Webb	3,485	270,465
Wilson	24,426	26,681
Zapata	7,318	98,980
TOTAL	728,247	2,461,088

SOURCE: A. W. Spaight, cited in Val W. Lehmann, *Forgotten Legions, Sheep in the Rio Grande Plain of Texas*, pp. 151–53.

NOTE: These counties encompassed the following, organized after 1882: Brooks, Jim Hogg, Jim Wells, Kenedy, Kleberg, Willacy, and Zavala.

* Later joined to Webb County.

Rocky Mountains. Millions of head were moved to Wyoming, Nevada, Colorado, Nebraska, Kansas, the Dakotas, Idaho, and Utah, where they crossed trails with vast herds of cattle and sheep coming from California. Other animals went to Corpus Christi or Indianola to be shipped overseas to Cuba and elsewhere. San Antonio replaced Corpus Christi as the center of the Texas wool trade. As San Antonio grew in prominence, drovers brought many flocks to the city for shipment by railroad to the Chicago meat packing plants. Sheepmen sent their wool to the city for sale and shipment to eastern markets.

The Rio Grande Plain range sheep industry reached a peak in the 1880s. In 1885 there were more than 3,500,000 sheep and 323,000 goats in the region. Afterward came a decline, slow at first and then increasing. The United States census showed 1,023,717 sheep in the Rio Grande Plain in 1890, 243,544 in 1900, and in 1910 only 110,168, with about half of them in the Hill Country portions of three northern counties.[19]

There were several reasons for the decline. Poor ranch management, including overgrazing, destroyed the range, turning the once lush South Texas grassland into the veritable desert of scrub brush and barren waste that it remains today. The practice of herding one sheep per acre proved too taxing for the buffalo grass and grama grass, hungry for rest and rain. Often absentee landlords allowed obnoxious and moisture-robbing mesquite, whose seeds were spread by the droppings of cattle and sheep, to grow unchecked. Bad weather, especially extensive drought in the early 1880s and the infamously cold winter of 1886, cut heavily into flock size. In 1893 a nationwide economic depression, possibly the worst of the nineteenth century, lowered prices for wool, eliminating or reducing many once profitable markets. All these factors and others were responsible. Today small flocks of sheep and goats are returning to South Texas, but for all practical purposes by the turn of the twentieth century "the Rio Grande Plain was devoid of sheep."[20]

For two decades the Rio Grande Plain had dominated the Texas range sheep and goat industry. The animals had been brought in from Mexico and the eastern United States, and crossbreeding to improve

[19]Cited in Evans, "Texas Agriculture," pp. 266–67.

[20]Lehmann, *Forgotten Legions*, p. 1.

wool quality had followed, with excellent results. The industry had helped to change the region from a largely unsettled and empty land to a place where livestock ranching predominated in every county. Sheep and goats, before they were carried to other sections of the state, had helped to make South Texas an economically viable region with firmly established herding traditions that were copied elsewhere during the Texas sheep boom.

5
The Texas Sheep Boom

EVEN while the mercurial Rio Grande Plain industry grew, forces were at work elsewhere in the state that created an expansive Texas sheep boom. Having started after the War with Mexico, the boom spread in the 1850s, hovered weakly or even retreated during the Civil War, and after the war started afresh. Rapid expansion came after the mid-1870s, when the entire area of the state west of the one hundredth meridian opened to sheepmen, who with cattlemen and others soon occupied state-owned land from San Antonio westward to the Pecos River and beyond. The boom peaked in the 1880s.[1]

A number of factors served to accelerate the boom. On a national level changing fashions in the 1850s caused men and women to favor wool garments over cotton. Then the Civil War cut the supply of cotton reaching New England mills, and textile manufacturers shrewdly converted their factories to wool processing. Many northerners turned away from cotton fabrics as an offensive reminder of slave labor, southern secession, and war. Likewise there was an improvement in the techniques of wool manufacturing. Carding, spinning, and weaving machines and the fulling process all underwent remarkable refinement as the wool textile industry strove to keep pace with international competition and with cotton manufacturing. The factory operators also began using new processes, especially the warp-dressing and cylindrical pressing machines. The warp-dressing procedure stood between spinning and weaving. It was a process by which yarn already spun was prepared for insertion in the loom as the warp threads (or the future longitudinal threads of the woven cloth). The process had previously

[1]See T. R. Havens, "Texas Sheep Boom," *West Texas Historical Association Year Book* 28 (1952): 3–17; Chester W. Wright, *Wool Growing and the Tariff; a Study in the Economic History of the United States*, pp. 252–54.

been a purely manual one. The roller pressing machine passed the cloth under heavy rollers, which were sometimes heated by steam, compressing the material to make it easier to handle. Despite the extraordinary technological advances, American wool-cloth manufacturing did not expand as it could have. It continued to lag behind cotton manufacturing, for instance, by about fifteen years, partly because of the greater difficulty in working the wool fiber and partly because of other causes. For one thing there remained a scarcity of fine wool for marketable cloth, although the spread of the Merino was helping overcome the difficulty. For another, the mass production by machinery of fine fulled cloth, a highly complex operation, posed a difficult problem that deterred the introduction of power-driven apparatus in the factories and the development of the factories themselves. The industry continued to experiment and by the late 1850s and 1860s had largely solved such problems. A vast western sheep boom, which included Texas, followed.

On a more local level also several factors influenced the boom. George W. Kendall's promotion of the Texas sheep industry was one. High wool prices during part of the Civil War and postwar demand for wool contributed to the sheep boom, too. The expanding Texas population was another cause. Americans pushed into Texas in large numbers. Most of them were traditional westward-moving pioneers. They left Tennessee, Kentucky, Alabama, or northern states to take up residence in Texas. Traders and storekeepers came in the vanguard, followed by a steady column of farmers. Dozens of new counties were laid out as people filled the state. In 1858 the editor of the *Northern Standard*, a newspaper at Clarksville, the northeastern gateway to Texas, declared that no fewer than fifty wagons of emigrants were passing through the town each day.[2] The pioneers crowded into eastern portions of the state and pushed out the frontier.

Under pressure to provide protection for the expanding frontier, the federal government completed the establishment of a line of military posts, generally a few miles in advance of the settlements, to deter Indians from raiding pioneers. Villages sprang up quickly. Soon additional military posts were established, ranging over a hundred miles

[2]Cited in Rupert N. Richardson, Ernest Wallace, and Adrian N. Anderson, *Texas, the Lone Star State*, pp. 181–82.

west of the settlements from the upper Brazos to the Rio Grande. Settlers wasted no time moving into the region. In 1860 the western extremity of settlement extended irregularly from Henrietta on the north through Belknap, Palo Pinto, Brownwood, Llano, and Kerrville to Uvalde.

East of the line the new arrivals accepted sheep. Their operations, however, remained small, the animals serving as supplemental income to cattle, cotton, or small-crop production. Usually the animals were kept in pens near the homestead. The farmer tended his own sheep and did his own shearing, lambing, docking, and castrating. The sheep in extreme East Texas were of a mean quality when compared with those of New England. They were of several varieties and mixed. Most of them had a little Merino blood. Few of the flocks exceeded 30 head, but their numbers increased during the boom. By 1880 the East Texas Interior Plains held 41,000 head, as did the East Texas Pine and Coastal plains. In the Black and Grand prairies the number totaled 183,000 head. Within another ten years the numbers of sheep in these areas had doubled.[3]

West of the 1860 line of settlement thousands of people arrived after the Civil War. Because cattle and sheep were plentiful and cheap there, most of the West Texas pioneers turned to farming and livestock production. The sheep were mainly *chaurros* crossed with a wide variety of American and European breeds, of which the Merino was the most popular. When the animals proved a good investment, many ranchers sold the wool clip, but they kept the spring lambs to expand their flocks. Flock sizes varied in the western regions but tended to number in the thousands and sometimes tens of thousands. Thus, the number of sheep in Texas increased, most rapidly in the early 1880s, as shown in table 2. The boom was statewide, but reached phenomenal proportions in the Rio Grande Plain and the Edwards Plateau.

Several immigrant groups participated in the Texas sheep boom—German, British, Mexican, and Basque among them. The Germans may have been the most important. Many of them came to Texas before the Civil War. During the war and the immediate postwar period some of them settled near San Antonio, dispersing through the Texas

[3]U.S. Bureau of the Census, *Tenth Census Report on the Productions of Agriculture, 1880* III, 170–72.

TABLE 2
Stock Sheep on Texas Farms and Ranches, 1867–87

Year	Number	Year	Number
1867	2,070,000	1877	2,896,000
1868	1,820,000	1878	3,186,000
1869	1,727,000	1879	3,505,000
1870	1,727,000	1880	3,715,000
1871	1,820,000	1881	4,230,000
1872	1,960,000	1882	4,864,000
1873	2,100,000	1883	6,200,000
1874	2,260,000	1884	6,600,000
1875	2,400,000	1885	6,620,000
1876	2,518,000	1886	5,675,000
		1887	5,150,000

SOURCE: Texas Crop and Livestock Reporting Service, *Texas Historic Livestock Statistics, 1867–1976*, p. 27.

Hill Country and up the Guadalupe River. They covered the area marked by present-day Comal, Kendall, Kerr, Gillespie, Blanco, Llano, Mason, Kimble, and San Saba counties west and north of San Antonio and by Fayette and Washington counties in the Schulenberg–La Grange–Brenham region. The Germans came in substantial numbers.[4] To accommodate the growing population, the German Aldersverein (Society for the Protection of German Immigrants) encouraged pioneers to consider Fredericksburg, established in 1846. Far out on the frontier, Fredericksburg after the Civil War attracted many German and other farmers who preferred running only a few sheep to large livestock operations. When they could get away from their farms, many of the farmer-settlers turned to freighting for supplemental income. In season they hauled wool and other freight between Hill Country villages, Kerrville, and San Antonio. Earning an enviable reputation in that occupation, some bought cattle and sheep with extra profits and slowly built sizeable herds and flocks.

Several of the Fredericksburg people moved into Mason County.

[4]Terry G. Jordan, *German Seed in Texas Soil*, pp. 31–59; Paul H. Carlson, "Texas Background: Spanish or American?" *West Texas Historical Association Year Book* 52 (1976): 66.

Among them, Louis Martin, the first sheriff of Gillespie County, settled in 1853 at Hedwig's Hill. In subsequent years three nephews and their families joined him. One nephew, Charles Martin, in 1858 erected a small store and with the help of his wife Anna operated it for several years. Catering to local ranchers, Anna Martin became well known as a sheep and cattle buyer. Eventually she controlled over twelve thousand acres of land in Gillespie County. Later in her career, with profits from her sheep and cattle ventures, she established the Commercial Bank in Mason, thus becoming one of the first female bankers in Texas.

Among German immigrants in the sheep industry, two of the most prominent were Charles Schreiner and Casper Real. Captain Charles Armand Schreiner, of French and German ancestry, came to America as a youth of fourteen, settling with his family in 1852 near San Antonio. Two years later he joined a company of Texas Rangers, in which he served with distinction. His mother's death in 1857 prompted his return to San Antonio. Near there he began ranching on a limited scale. Shortly afterward he moved to a new ranch along Turtle Creek in Kerr County near Kerrville, a region inhabited by Indians and cattle. The cattle were so cheap, reminisced Schreiner, "that people would not go to the trouble to brand the calves, and in some parts of Texas the maverick yearlings were killed for their hides."[5] He turned to wool growing.

During the Civil War, he left behind his ranch, a new wife, and a young child to fight for the Confederacy, serving with the Third Texas Infantry. Mustered out of the army in San Antonio at the close of the war, Schreiner walked seventy-five miles to his ranch on Turtle Creek rather than spend his last gold piece for transportation.

Back on the ranch Schreiner returned to his neglected sheep. Using the Spanish organizational system of range sheep husbandry, he built his band to forty thousand head, breeding Delaines, so well adapted to the brushy, mountainous country near Kerrville, to *chaurros*. Eventually, his ranch included some 500,000 acres in Edwards, Kerr, and Kimble counties.[6] In the meantime, however, he had moved

[5]Quoted in W. Eugene Hollon, "Captain Charles Schreiner," *Southwestern Historical Quarterly* 47 (October, 1944): 153.

[6]Edward N. Wentworth, "The Golden Fleece in Texas," *Sheep and Goat Raiser* 24 (December, 1943): 27; "A Patriarch Passes to His Father," *Sheep and Goat Raisers' Maga-*

to town, which boasted a business district of two or three saloons and a sawmill. Here, Schreiner established a general merchandise store. This humble beginning soon developed into a great diversity of enterprises, including a busy bank. Out of the banking operation came one of Captain Schreiner's most influential actions. Whenever a cattleman of the upper Guadalupe Valley area entered Schreiner's bank to borrow money, the rancher found it necessary to use at least half of the loan for sheep raising. Although sheep were resented for several reasons—mainly because they were too much trouble, requiring constant attention—many borrowers agreed to Schreiner's unusual terms. The practice spread sheep husbandry rapidly.

Casper Real, Schreiner's brother-in-law, was another distinguished German sheepman. In 1859 Real brought the first flock to Kerr County and settled along Turtle Creek about six miles from Kerrville. Neither his land holdings nor his flocks were as large as Schreiner's, but he paid exceptional attention to breeding up his animals, importing fine Delaines from Ohio. The sheep arrived by wagon from Indianola, where they had been taken by steamer. Real, who maintained both purebreds and crossbreds based on Mexican *chaurro* blood, sold many of his sheep to cattlemen who had borrowed money from Schreiner. Some years later one of Real's sons, Robert, took over management of Schreiner's sheep operations while another, Arthur, carried forward Real's ranch.

In part because of the Schreiner-Real interests, German settlers developed a combination sheep- and cattle-ranching industry in the Hill Country. Later the practice characterized Anglo ranching operations. This approach tended to reduce financial risks that often plagued ranchers who depended upon one type of livestock production. According to the federal census of 1850, one-tenth of all Germans in the western counties owned sheep, and by the 1870s many of them ran more than a thousand head.[7] As a result of this combination sheep and cattle ranching no animosity developed between the sheep ranchers and cattle ranchers of the Hill Country, as it did in many other parts of the West.

---

zine 7 (February, 1927): 3; W. E. Blanton, "The Life of Captain Charles Schreiner Is an Inspiring Memory," *Sheep and Goat Raisers' Magazine* 7 (July, 1927): 9–11.

[7]Cited in Jordan, *German Seed*, pp. 149–50.

British immigrants also participated in the expanding Texas sheep industry. One group of them were "remittance men," young men of upper- and middle-class families, unable to find work in areas usually open to their class, such as the learned professions, the army or navy, or the civil service. They came to America, where they, writes one historian, "ransacked our entire West for their fame and fortune." They received regular allowances from their families in England, Scotland, and Ireland. In Texas their role in the sheep industry was as interesting as it was noteworthy. After the Civil War many Texans, attracted by the exciting profits that could be made selling beef, took over cattle ranching. Cowboys sitting high in their saddles rode quickly about the range on horseback with tight reins on wild steeds and a proud sense of self-importance about them. The sheepherder usually worked on foot, or, if mounted, rode quietly with a loose rein and a gentle horse. Since the contrast tended to damn the sheepman's occupation, few Texans of the era were willing to handle sheep. But British immigrants, especially the remittance men, proved impervious to such social opinions, for as Winifred Kupper, whose family was English, claims, the black condemnation of the sheepmen held no terror for the Englishman "because, after all, was he not an Englishman, and didn't he feel a tinge of pity for anyone who wasn't?"[8]

There were many such self-assured Britishers arriving in Texas after the Civil War. They provided respect for sheepherding. They unhesitatingly offered to herd sheep on shares, hired themselves out as herders, or invested in animals and land when they had enough financial reserves. They scattered over a wide area of Texas as the sheep industry spread through the state.[9] Some succeeded. Others went back to England disillusioned and broke.

Not all the Britishers were remittance men. Thomas Hughes from England settled near Boerne in the Hill Country. There he raised thousands of sheep, combining the sheep-raising cultures of the British and the Spanish. Sandy McNubbin, from Scotland, settled in Nueces County after the Civil War. He herded sheep on shares and became an

[8]Wentworth, "The Golden Fleece," p. 26; Winifred Kupper, "Folk Characters of the Sheep Industry," *Cattleman* 26 (March, 1940): 89–90; Winifred Kupper, *The Golden Hoof: The Story of the Sheep of the Southwest*, p. 65.

[9]See James I. Fenton, "Big Spring's Amazing Tenderfoot: The Earl of Aylesford," *West Texas Historical Association Year Book* 55 (1979): 135–48.

expert shearer before settling in Corpus Christi. W. S. Sutherland from Scotland herded sheep in the Rio Grande Plain. The Adams brothers, starting with only a small band of animals, amassed a fortune in sheep. Later they were responsible with others for the formation of Jim Wells County. Jim Shannon, from New Zealand, who began by herding sheep, became a champion shearer and quietly collected one of the largest fortunes in Texas. He surprised everyone when he turned up at a San Angelo wool warehouse wanting enough sacks for ten thousand pounds of wool. Later he established the J. M. Shannon Company, a sheep and wool enterprise with holdings in several West Texas counties and New Mexico. Captain Charles Gordon, a retired British military officer, settled on a sheep ranch near Kerrville. He kept his ranch open to Englishmen who wanted to learn sheep husbandry. Many a remittance man came to the ranch equipped for the herder's life "with a plentiful supply of hand-embroidered sheets, handsome dresser scarves, and antimacassars for proper embellishment of his room."[10] In 1882 Robert Maudslay from England stopped at Bandera. From there for more than two decades afterward he trailed sheep over much of the Southwest.

The British were often well educated and brought their books and learning with them to the sheep camps. Much of what is known today about early Texas herding came from the diaries, letters, journals, and memoirs of learned British sheepherders. They also introduced English mutton sheep, like Lincolns and Cotswolds. The Lincoln (Lincolnshire) originated in flat, marshy Lincoln County, England. This huge mutton variety is said to be the heaviest breed in the world. Rams weigh 250 to 350 pounds. They were imported to increase the body size of Texas animals. Cotswolds are slightly smaller, but more active, than their cousins, the Lincolns. A long-wool mountain breed of the Cotswold Hills in England, they were bred to Spanish ewes on the Texas ranges.

Basques from the Pyrenees Mountains also came to Texas during the sheep boom. Originally coming to America in the days of the California gold rush, they soon shifted to sheep, herding over much of the West. Some of them drifted into Texas during the late stages of the boom. They universally accepted sheep on shares. The successful ones

[10] Kupper, *Golden Hoof*, p. 69.

among them accumulated large enough bands to enter business for themselves, grazing the sheep wherever they could—often on public land claimed by others. But with limited English and with clannish ways, they were seldom very popular. Americans tended to look down upon them as a filthy, treacherous, and meddlesome bunch. But they were excellent and hardworking sheepmen. Today their descendants dominate many sheep-raising counties of the West.

Whether immigrant or American, before the late 1890s sheepmen found little trouble in securing grazing land. This was the era of "free grass." West Texas was an empty land with plenty of good grass, and who held legal possession of it was not important. In the early years most herders thought it was foolish to buy or lease property. As conditions on the range became more crowded, some ranchers began to buy land. Others leased property at about four cents per acre yearly from railroads or from state-held public-school lands. Up to January 1, 1885, the school land board leased 1,802,805 acres at an average price of seven cents an acre. At that time, too, inspectors for the state found many cases in which school land was fenced and sheep and cattle were grazing on it, but no lease had been obtained. The state disposed of much of its land through grants as headrights, bounties, donations, and homesteads. The state also sold millions of acres, but at prices well below their true value. The public-school lands, for example, were sold before 1900 at an average price of $1.64 per acre.[11]

With land plentiful and cheap, Texas sheepmen of whatever background fanned out to cover all parts of the state. As they did so, the number of sheep increased. While reports on the number and value of sheep in Texas vary considerably, there were about 100,000 head in 1850, more than 680,000 ten years later, and in 1870 over 1,200,000, placing Texas eleventh among states in numbers of sheep.[12] Five years later Texas had advanced to fifth place. The increase continued year by

[11] Thomas Lloyd Miller, *The Public Lands of Texas, 1519–1970*, pp. viii, 187–89, 248.

[12] For examples of differences in estimates, see Texas Crop and Livestock Reporting Service, *Texas Historic Livestock Statistics, 1867–1976*, p. 27; *Texas Almanac for 1980–81*, p. 578; Havens, "Texas Sheep Boom," p. 11; U.S. Bureau of Animal Husbandry, Fourteenth Annual Report, p. 297; Samuel Lee Evans, "Texas Agriculture, 1880–1930" (Ph.D. diss., University of Texas, Austin, 1960), pp. 266–67; Wright, *Wool Growing and the Tariff*, pp. 252–54, 271.

TABLE 3
Sheep in Texas, by Region, 1880 and 1890

Region	1880	1890
East Texas Interior	41,000	80,000
East Texas Pine and Coastal Plains	41,000	76,000
Black and Grand Prairies	183,000	346,000
South Texas Prairies	317,000	207,000
South Texas Plains	1,222,000	746,000
Western Cross Timbers	108,000	384,000
Red River Rolling Plains	35,000	175,000
Edwards Plateau	376,000	1,295,000
Northern High Plains	78,000	12,000
Southern High Plains	—	73,000
Stockton–Trans-Pecos Mountains and Plateaus	10,000	61,000
TOTAL	2,411,000	3,455,000

SOURCE: Samuel Lee Evans, "Texas Agriculture, 1880–1930" (Ph.D. diss., University of Texas, Austin, 1960), p. 267.

year. In 1885 the total reached 6,620,000. Another report showed Texas with 4,761,831 head in 1886, ranking second only to California in numbers of sheep.[13] The distribution of sheep by region in Texas in 1880 and 1890 is shown in table 3.

Concomitant with the increase in numbers of sheep and lambs, Texas expanded its wool production. The large output came in part from the presence of additional animals, but more important was the

[13]Texas Crop and Livestock Reporting Service, *Historic Livestock Statistics*, p. 27; Report of the Commissioner of Agriculture, 1886, p. 405, cited in Havens, "Texas Sheep Boom," p. 11. Using reports from the Tenth U.S. Census, 1880, Samuel Evans, in his study of Texas livestock, has presented statistics that probably underestimate the numbers (Evans, "Texas Agriculture," pp. 266–67). The U.S. Bureau of Animal Husbandry's Fourteenth Annual Report in 1891 (p. 297) listed Texas as having had 7,952,275 head of sheep in 1883.

dramatic growth in the production of wool per sheep. Improved breeding showed profitable results. In 1850 when the Spanish *chaurro* dominated the ranges, Texans averaged less than 1½ pounds of wool per animal. In 1860 Texas sheep produced a wool clip of 1,500,000 pounds, representing an average of nearly 2 pounds per animal. The average increase in wool poundage per animal continued steadily upward during the remainder of the century, reaching an average of over 5 pounds per clipped sheep. Some wool growers reported exceptional results. Colonel John James, owner of a large spread west of San Antonio, observed in the 1870s that his sheep produced an average wool clip of 5 pounds per year at thirty cents per pound, a gross profit of $1.50 per animal. This, combined with an average lamb crop of 80 percent of pregnancies from several thousand ewes bred to Merino rams, created a sizeable return at minimum expense. In 1884 the flocks of the Honorable H. J. Chamberlin of Milam County, numbering twelve hundred head, yielded 10½ pounds of wool per head, with pureblood rams running from 15 to 33 pounds. Total production increased, too, reaching 14,917,068 pounds in 1890.[14]

Shearing the wool from sheep in regions north of San Antonio occurred once a year, beginning in April or May, but otherwise did not differ from the methods employed in the Rio Grande Plain. Mexican crews, receiving three to four cents a fleece, worked at a covered platform on which the sheep were thrown down, tied, and shorn while the flockmaster and his assistants tallied, tied, and sacked each fleece. The dry flocks were shorn first, the suckling ewes last to avoid any loss from separating the ewes and lambs too early.

Freighters hauled the wool to markets. Many people owning horses and wagons designed for the purpose developed a special trade in wool freighting. These teamsters, whose system differed from the Mexican cart trade, often hitched three to four heavily loaded wagons, attached one behind the other, to five or more pair of horses. Economy of time and money demanded large loads, but as one side effect many drivers overloaded wagons. The practice resulted in overturned loads, broken axles, lost wool sacks, and other difficulties. On the narrow, twisting roads characteristic of the Hill Country, freighters heading in

[14]Report of the Secretary of Agriculture, 1893, cited in Havens, "Texas Sheep Boom," p. 13. See also *Texas Almanac*, 1980–81, p. 578.

opposite directions occasionally met. Problems followed, for one of the freighters must back his team down to a wide spot in the road. Some obstinate teamsters were known to unhitch their animals and camp out, waiting for the other driver to give in. Heavily loaded wool wagons were a common sight at sheep-shearing time.

In the 1870s Colonel Thomas Clayton Frost dominated the Texas wool trade. At the close of the Civil War, Frost purchased a train of wagons with teams and began to haul freight, including wool, between San Antonio and the Gulf coast at Corpus Christi and Indianola. In 1868 with his brother John he opened a general merchandise store, known as T. C. Frost Company, in the main plaza in San Antonio. The store's operators quickly won the confidence of farmers and ranchers in the area. In exchange for credit at the store, Frost secured the right to market the wool clip of area sheepmen. It proved a profitable business that expanded yearly. Frost acted as a warehouseman and a commission salesman and in the 1880s may have handled more wool than any other man in the country. He soon built several large warehouses to store the clip. Twice a year Frost gathered local wool (in extreme South Texas wool growers sheared their animals twice yearly), which he kept until favorable marketing conditions prevailed.[15]

Colonel Frost kept the resulting income on deposit at his store. Between sales the growers drew against the deposit for supplies of all kinds. From this practice, today's Frost National Bank of San Antonio emerged. At first Frost operated the bank in one small office at the back of the merchandise store, but in the 1880s, when activity became too great for the store, he constructed a separate building to handle the banking operations.[16] In 1892 he abandoned his wool warehousing business.

Captain Schreiner of Kerrville also pioneered wool warehousing. From the 1860s, when sheep were first brought into Kerr County, he bought and sold wool. For several years he freighted wool by wagon through San Antonio to the Gulf coast. But as railroads pushed westward through Texas, Schreiner moved to secure trackage in his town. He rallied the townspeople to provide financial support for a railroad

[15] *San Antonio Light*, June 20, 1954; Edward N. Wentworth, *America's Sheep Trails*, pp. 611–12.

[16] *San Antonio Light*, June 20, 1954.

and added $150,000 of his own money. Using the funds to purchase land for sidings and right-of-way, he and Kerrville citizens induced the San Antonio and Aransas Pass Railroad to extend its line there. In 1887, when the railway reached his city, Schreiner erected on the main track a warehouse that handled wool from several counties in the Hill Country north and west of Kerrville. In 1890 he doubled his warehouse capacity and eight years later erected a large stone warehouse on a convenient railroad siding.[17] By this time he had become the largest wool warehouseman in Texas.

Schreiner and Frost sold their wool to scattered buyers. Some wool they shipped to Mexico, some they kept for tiny local markets, and most of the yearly clip they delivered to New England. By the 1870s Boston had reached a position of world prominence in the wool industry, and its agents annually purchased wool in many parts of the state. In 1870 buyers representing more than a thousand New England firms posted prices in Texas. Afterward the number declined somewhat, as wool warehousing concentrated in cities like San Antonio, Encinal, and Kerrville before railroad extension again allowed additional facilities. Warehouses, too, began to appear in most West Texas towns.

But the expanding Texas sheep industry was not without its difficulties. The prices agents for New England wool manufacturers posted were on the basis of "washed wool," a fleece that had been cleaned before shearing. But the Spanish system of sheep husbandry practiced by most wool growers in Texas never included washing before shearing. Water was too scant and the numbers of sheep were too great in relation to the number of men employed to make the practice profitable. Also, daily extremes of temperatures in Texas were wide enough to risk chilling the wet sheep when evening winds struck upland bed grounds. Moreover, washing represented an essential phase of sheep operations only as long as spinning and weaving remained a home industry. When spinning and weaving became commercial, as they did in the 1860s, or even when local scouring mills were established, the need for washing ceased. Nevertheless, in the 1860s and 1870s washed wools sold at a premium, and Texans found that their unwashed wools were penalized. They felt such pricing was unjust.

[17] Hollon, "Captain Charles Schreiner," pp. 158–59; Blanton, "Life of Captain Schreiner," pp. 9–11; "Patriarch Passes," p. 3.

A related grievance resulted from the New England manufacturers' desire for foreign wool. Claiming a public demand for fabrics from types of wool not grown in the United States, the manufacturers imported cheap wool, thereby depressing the American market. Although Americans had gained some import protection through tariff duties since the landmark protective law of 1816, Texas wool growers, desiring high tariffs, blamed manufacturers' support of low tariffs when wool prices dropped. A mutual distrust, complicated by sectional jealousies and misunderstandings, resulted. Indeed, the lack of harmony between wool growers and wool manufacturers in the United States helped to create the National Wool Growers Association, established in 1865.

In that year, according to the United States Revenue Commission created by Congress to make a study of the country's revenue system, the wool industry was one of the nation's most important sources of internal revenue, exceeded only by the liquor, iron and steel, and tobacco industries. One of the commission's three members, Stephen Colwell, a Pennsylvania lawyer and iron manufacturer, assigned by commission chairman David A. Wells to investigate the wool industry, indicated that there had never been a full supply of the wool needed for health and comfort in America and that present production rates ought to be doubled. "Sheep husbandry," he said in 1865, "should be stimulated and promoted until our supplies of wool shall exceed 200,000 pounds."[18] To achieve such goals the commission proposed to bring the wool growers and manufacturers together. It wanted them to consider a new tariff measure as well as all other matters of mutual interest.

The commission was successful. The wool manufacturers had a national organization in 1864. Through their group, the National Association of Wool Manufacturers, the commission in November and early December, 1865, contacted representatives of several state and local wool growers fraternities, urging attendance at a conference in Syracuse, New York, to discuss mutual concerns respecting wool producing and manufacturing. In response, some two dozen wool-growing

[18] "History of the National Wool Growers Association," *Southwestern Sheep and Goat Raiser* 7 (December, 1936): 28–30, 57–59; Stephen P. Colwell, "Special Report No. 3," May, 1866, presented to Secretary of the Treasury, as cited in Irene Young et al., "Men, Sheep and 100 Years," *National Wool Grower* 55 (January, 1965): 2.

representatives from several states, but not Texas, met December 12, the day before the scheduled conference with the manufacturers. J. W. Colburn, president of the Vermont Wool Growers Association, presided at the meeting, and William F. Greer of Ohio served as secretary. Texas, having no state organization, was not represented. Indeed, most Texas wool producers probably knew nothing of the meeting. However, Henry S. Randall, president of the New York Wool Growers Association, who was familiar with Texas wool problems, knew about the concerns of Texas sheepmen. Randall reported at some length on the purpose of the wool growers' preconference meeting, saying it was to take into consideration the propriety of creating a national group corresponding to the national organization of manufacturers. Before the evening ended, the representatives had created the National Wool Growers Association.[19]

The following day, December 13, 1865, the wool growers and manufacturers met. Randall, just elected national president of the wool growers, presided. He led the representatives through discussions of such items as washed versus unwashed wool, the Merino ram versus other breeds for upgrading wool stock, and improved manufacturing equipment. But for both groups the immediate concern was to secure adequate protection in a new tariff planned by Congress but still under consideration by the revenue commission. Neither group was satisfied with the rates then in effect, but the manufacturers, led by Erastus Bigelow, their president, and John L. Hayes, their secretary, also wanted protection from the unsettled state of the nation's currency. Hoping for resumption of specie payments for "greenbacks" (unbacked paper money printed to finance the Civil War), the manufacturers bought wool on a day-to-day basis. The growers feared resumption. They wanted unbacked paper money to float, thus maintaining high wool prices.

The tariff, therefore, dominated the conference. In fact, there is some evidence to suggest that Bigelow, a successful carpet manufacturer, and a few energetic manufacturers from New England who produced chiefly carpets, blankets, and worsted goods engineered the meeting to their benefit. In order to meet foreign competition, they

[19] Frank W. Taussig, *The Tariff History of the United States*, pp. 198–99; *Texas Almanac*, 1861, pp. 160–66.

agreed to the high duties on raw wool that the producers sought in exchange for a complicated compensation system that would always get the manufacturer his wool as though it were free of duty. They also sought lower rates for carpet wools than for the better grades. Apparently knowing about Bigelow's schemes, many woolen manufacturers refused to attend the conference.[20]

The growers and manufacturers at the Syracuse conference agreed to prepare a tariff resolution. Executive committees of each group met with one another for several weeks before presenting a joint report on the tariff to the revenue commission on February 9, 1866. Some weeks later each committee made out a tariff schedule for its own industry. The revenue commission, accepting the joint report, recommended a tariff bill to Congress. The measure, except for one slight change made by Congress, became law on March 2, 1867.

This law, the Wool and Woolens Act of 1867, eventually became one of the most important tariff acts in the history of duties on wool and woolens. It classified imported wool into three groups: clothing wools, combing wools, and carpet wools (see appendix A). Clothing and combing wools enjoyed similar rates: when value of the fleece equaled thirty-two cents or less per pound, the duty was ten cents per pound plus 11 percent ad valorem; when value exceeded thirty-two cents, duty would be twelve cents plus 11 percent ad valorem. Clothing wools if imported washed (cleaned) paid double duty. Duties on the carpet wools stood at three cents per pound when the value was twelve cents or less per pound, six cents when the value exceeded twelve cents.[21]

The wool tariff rates, not substantially changed until 1894, had little effect on the Texas sheep boom. Growers in the state, receiving some protection from foreign competition, found their own wool more

[20]Taussig, *Tariff History*, pp. 195–97, 199; Harry J. Brown, "The Fleece and the Loom: Wool Growers and Wool Manufacturers during the Civil War Decade," *Business History Review* 29 (March, 1955): 1–27. Wool manufacturers estimated that it took about four pounds of fine raw wool to make one pound of cloth. Also a fine Merino fleece will shrink 50 to 75 percent in scouring, a quarter-blood will lose only about 25 to 35 percent. Products of low value were manufactured, by and large, from wools of low grade or from mixtures of wool with cotton or shoddy. The compensation required on such wool would be below the four-to-one ratio. The carpet manufacturers, therefore, would receive significantly higher compensation than the woolen manufacturers (Arthur Harrison Cole, *The American Wool Manufacture* II, 6–7, 14–15).

[21]Brown, "Fleece and Loom," pp. 1–27; Karl Everett Ashburn, "Tariffs and Wool Duties since 1867," *Sheep and Goat Raisers' Magazine* 10 (November, 1929): 104.

in demand, but for two years immediately following the tariff a brief slump hit the sheep industry. Eastern buyers, who had bought heavily just before passage of the bill, had huge supplies on hand. Moreover, at the end of the Civil War, the government dumped its surplus of woolen goods, no longer needed for its armies, onto the market, and cotton production began to increase again. Prices for wool, although remaining relatively high for the nineteenth century, declined from more than thirty cents per pound in 1867 to 23.1 cents per pound in 1880. Partly as a result, the Texas sheep industry contracted briefly after the Civil War before resuming its remarkable expansion.[22]

The relationship between Texas wool producers and the National Wool Growers Association was informal in the late nineteenth century. Texas had no state organization to coordinate its efforts and concerns until 1915, and the national group held only periodic national conventions until 1901, when regular annual meetings were begun. Nevertheless, Texans occasionally attended the national meetings to represent themselves or local—county—associations. William L. Black, for example, attended the 1885 meeting of the National Wool Growers Association in Saint Louis, his former home and business headquarters.

Despite the nonaffiliation of Texans, Henry S. Randall, having written about the Texas industry, was familiar with Texas sheep and wool production. Randall was president of the National Wool Growers Association from its founding in 1865 to 1876. While the tariff remained the major topic for the organization, the National Wool Growers Association dealt with a wide spectrum of concerns, including washed versus unwashed wool. This problem was of no little importance to Texans, whose wools were unwashed. It was customary for buyers to lower the price by as much as one-third on unwashed wools. But once Texans learned, through the National Wool Growers Association, that manufacturers had no objection to buying unwashed wools, growers were able to disregard or withstand the demands of the buyers to lower the price.

As Texas sheep raising expanded again in the 1870s, the Angora goat and mohair industry picked up, if slowly. After W. W. Haupt and Robert Williamson introduced Angoras to Texas, others began brows-

[22]Texas Crop and Livestock Reporting Service, *Historic Livestock Statistics*, pp. 37–39. See also Wentworth, *America's Sheep Trails*, p. 394; and *Texas Almanac for 1980–81*, p. 578.

ing the animals. Judge J. P. Devine, who brought Angoras to his ranch near San Antonio in the 1860s, was among the first. Devine, having acquired the goats from Colonel Peters in Georgia, took much pride in the animals. He maintained pure-blooded stock but also bred some of the billies to common Spanish does. He enjoyed surprising success at increasing mohair clips. In a letter to John L. Hayes, one of the first chronicles of the mohair industry in America, Devine indicated that he had clipped much heavier fleeces in Texas from goats that had browsed in Georgia. The reasons for the good fortune, he explained, were the inexhaustible quantity of rich evergreen food throughout the winter and the dry atmosphere. His large fleeces, which averaged from four to five pounds on crossbred animals, encouraged many stockmen to adopt Angoras.[23]

Colonel W. D. Parish of Seguin purchased and brought to Texas from New England purebred Angoras. Like Devine, he bred them to Spanish goats with good results. Later Parish took his small band in wagons to Kendall County, where they could browse in the dry, brushy upland range of that region. A contemporary of Parish, George W. Baylor of Uvalde County, wrote that in his opinion "Mr. Parish did as much towards starting the Angora goat industry in Texas as any one."[24]

There were many other early breeders. The Arnold brothers and J. V. Abrams of Frio were among them. The Reverend D. S. Babb ran a flock near Sonora in Sutton County. R. H. Lowry of Camp San Saba, one of the first Angora raisers in Texas, started with goats purchased from W. W. Haupt. William M. Landrum and his sons, W. E. and Frank Landrum, raised Angoras in Uvalde County. Charles Schreiner of Kerrville, realizing the potential importance of the silky white goat, was instrumental in establishing Angoras in the Hill Country. As he had done with sheep, he encouraged area ranchers to invest in mohair goats to diversify their ranch livestock.

Since it was soon found that Angoras throve best in rocky, rolling hills, ranchers in the Edwards Plateau were among the first to raise the goats. There the number of Angoras increased slowly until after 1880, when expansion came more rapidly. Individual flocks remained small, too, in the early stages of the business, seldom numbering more than

[23] Quoted in William Leslie Black, *A New Industry—Or Raising the Angora Goat*, p. 77.

[24] Ibid., p. 78.

two hundred head. But Angora does, by often giving birth to twins, made rapid flock expansion common.

Mohair markets also improved. As early as 1863 textile manufacturers began installing proper machinery. But since it was expensive, New England mills were slow to incorporate the equipment. Nevertheless, when they found that mohair fibers attracted considerable attention among buyers of their goods, they moved to adapt their textile machinery. By 1880 mohair had become a popular fiber, used in a variety of fabrics. Manufacturers of railroad passenger cars found it especially attractive for plush upholstery.

After 1880 changes came to the Texas sheep and goat industry—one wrought by the westward-moving population. The number of people living west of an irregular line drawn from the common boundary of Cooke and Montague counties north of Fort Worth on the Red River southward to Bandera and thence to the Gulf a few miles south of Corpus Christi totaled about 165,000 in 1880. In 1900 the number reached half a million. People had streamed in as railroads extended their tracks into the vast area, driving the range sheep and goat industry westward before them.

In the region east of the line the sheep industry underwent consolidation and reorganization. The open-range and Spanish systems of management disappeared there as farmers adopted small flocks to raise wool for income to supplement other endeavors. However, to provide mutton for such distant markets as Kansas City and Chicago, many additional farmers for the first time took to raising sheep. As a result, for ten years after 1880 nearly every section of the state east of the line found its sheep population increasing, but there were few large-scale operations. Farmers in extreme East Texas nearly doubled their sheep numbers. The same was true of farmers and ranchers on the Blackland Prairies and in the Cross Timbers. But in the 1890s the numbers of sheep east of the line declined sharply.

The region west of the line was isolated and sparsely populated. Only a few farm operations existed there. Instead livestock men herded their cattle, sheep, and goats on immense, empty ranges in the region. They adopted ranching operations patterned on old Spanish systems adapted to meet their needs. As settlement spread, the numbers of sheep increased until the mid-1880s.

By then the Texas sheep boom was over. During the boom the

Texas sheep and wool industry had undergone a remarkable transformation. It had moved steadily westward over a wide area. It had adopted improved breeds with good results. It had developed new marketing facilities in wool-growing areas. It had provided employment for thousands of people as herders, shearers, freighters, or warehousemen. It had attracted American settlers and European immigrants to the land. In short, the Texas sheep boom had been good for the state. It had helped to make Texas a leading sheep-producing area in the nation.

As sheep and wool production expanded westward in Texas, Panhandle *pastores* entered the state eastward from New Mexico.

6
The Panhandle Pastores

As the sheep boom swept across the state's southwestern regions, similar inducements brought New Mexican sheepherders with their animals eastward into the Texas Panhandle. For perhaps a decade or less, from about 1874 to 1884, sheep covered the Canadian River Valley, but the timid animals sometimes grazed as far south on the Llano Estacado as Tahoka Lake and as far north as the present Oklahoma border. In the vast, empty tableland the sheepmen developed an unhurried and uncomplicated, but rugged, agrarian life-style. They brought with them many New Mexican modes of living and methods of sheep raising. Their romantic ventures proved short lived, yet for Panhandle agriculture the *pastores* played a salient role.

The Canadian River Valley attracted most of the *pastores*. In 1874 the river presented a beautiful tableau. Hardly more than twenty feet wide, it had no sand bars but was deep with clear, living water. Its banks along nearly the entire route through the Panhandle were fringed with fruit bushes and trees, including wild chokeberries, gooseberries, grapes, and plums. Stands of cottonwood trees hugged the banks as well. One could find sun perch in the small arroyos of the river and large channel catfish in the Canadian. Buffalo, deer, and upland game birds visited the stream. Until trappers killed the animals, beaver and beaver dams were numerous. As the dams deteriorated, the water ran unchecked, altering the nature of the river, and cattle, when they came, destroyed the bushes.

No one knows when the first sheep came to the Panhandle, but certainly it was long before any permanent settlers arrived. The sheepherders came from New Mexico, where flocks were numerous and large by 1874, overrunning the available range. Sheepmen turned their eyes to Indian land first as they sought to expand their grazing circuits and then to the Texas Panhandle with its enormous and beckoning Llano Estacado, the Staked Plains, containing a veritable ocean of shimmering grass. Cut by the Canadian, Red, and other rivers, the northern reaches of this high plains country lured *pastores* to its lush valleys and endless pastures.

Indians, however, temporarily blocked penetration into the Panhandle. Comanche, Kiowa, and southern Cheyenne hunted buffalo in the region and, before being confined to reservations in present-day Oklahoma, camped in the canyons and ravines that protrude deep into the Llano Estacado. After accepting reservations, the Indians enjoyed hunting and trading rights in the Panhandle, and they resented the encroachment of any group into their domain, occasionally using diabolical means to discourage *entradas*. The *pastores*, simple, poorly armed men, made easy targets for the vengeful Southern Plains warriors.

Nevertheless, a few daring sheepmen tested the Indians. During the 1860s a group of New Mexican families founded a settlement along the Canadian River below Parker Creek in present Oldham County but abandoned the site shortly afterward. An old sheepherder named Antonio Baca claimed to have grazed thirty thousand head of sheep in No Man's Land, the present Oklahoma Panhandle, soon after the Civil War.[1] No doubt some of the earliest *pastores* entered the Panhandle as they extended their summer grazing circuits onto open range before driving their sheep back to winter pastures in New Mexico.

In any event, documentary evidence supports the view that New Mexicans, other than *comancheros* (men who traded with Plains Indians, especially Comanches) or *ciboleros* ("buffalo hunters"), lived permanently in the region no earlier than the end of the Red River War, fought by the United States Army against the Plains tribes in the Panhandle during the fall and winter of 1874–75.[2] After the war, however,

[1]John L. McCarty, *Maverick Town, the Site of Old Tascosa*, p. 8; John Arnot, "A History of Tascosa," typescript, Earl Vandale Collection, Archives, University of Texas, Austin, pp. 4–6; Oscar A. Kinchen, "Pioneers of No Man's Land," *West Texas Historical Association Year Book* 18 (1942): 24–25. See also A. J. Taylor, "*Pastores* in the Texas Panhandle," ed. Robert E. Simmons, *Hale County History* 10 (February, 1980): 1–46.

[2]Frederick W. Rathjen, *The Texas Panhandle Frontier*, pp. 101–103; Charles L. Kenner, *A History of New Mexico–Plains Indians Relations*, pp. 116–17. Some members of the staff at the Panhandle-Plains Historical Museum in Canyon, Texas, are, nevertheless, convinced that *pastores*, on their yearly circuits from New Mexico, entered what is now the Texas Panhandle long before the Civil War. Their energetic investigations may

Indians continued to wander into the Canadian River Valley. They came down the river from reservations in New Mexico or up the river from reservations in Indian Territory (Oklahoma), hunting in season the few remaining buffalo and trading with the *pastores*. In some ways the trading represented an extension of the *comanchero* business. *Pastores* provided cattle and trade goods from Las Vegas, New Mexico, for horses and a few other items the Indians could offer. As the Panhandle population grew and American hide hunters wiped out remnants of the last herds of the magnificent brown buffalo, the Indians came for trade less frequently.

The first important group of *pastores* in the Panhandle came with Casimero Romero. A former *comanchero* who had traded on the Llano Estacado and had long been a resident of Mora County, New Mexico, Romero in November, 1876, led a large party—including perhaps a dozen Romero employees and three friends, all with their families—eastward to settle in the Canadian River Valley at the site where Tascosa later developed. A man of substance, Romero and his family traveled in a fine, enclosed carriage. Behind the carriage trailed twelve or fourteen freight wagons—large schooners bought from the United States Army—each pulled by four yoke of oxen. The wagons carried household goods, lumber, ranch equipment, and miscellaneous supplies to last for a year. Following the caravan were about 3,000 head of sheep belonging to Romero and another 1,500 head belonging to Agapito Sandoval, one of Romero's friends; several horses; and enough cattle to provide beef and milk for the settlers.[3]

After reaching its destination at a bend in the Canadian near its junction with Rica Creek, the group prepared a winter camp. Here, inside a stand of cottonwood trees, it drew its wagons into a circle. After setting a tent in the enclosed area, the pioneers had a comfortable place in which to live until there was time to build permanent

provide new information. For historians of the Panhandle and the Texas livestock industry, their work in this area is important. Dugouts built by New Mexican sheepherders have been found over much of the Texas High Plains (see David J. Murrah, *C. C. Slaughter: Rancher, Banker, Baptist*, pp. 34, 44).

[3] Roy Riddle, "Casimero Romero Reigned as Benevolent Don in Brief Pastoral Era," *Amarillo Sunday Globe-News*, Golden Anniversary Edition, August 14, 1938, Section C, p. 29; Jose Ynocencio Romero, "Spanish Sheepmen on the Canadian at Old Tascosa," ed. Ernest R. Archambeau, *Panhandle-Plains Historical Review* 19 (1946): 46–48; Edward N. Wentworth, *America's Sheep Trails*, pp. 118–20.

homes—*plazas*—of adobe, sun-dried brick and mud. The men rode about the area, choosing good locations for future homes. Each family selected an area near a fresh, flowing spring. Romero chose a site near Atascosa Creek, later shortened to Tascosa Creek, along the banks of the Canadian River. The spot was near several large springs that attracted hundreds of buffalo. Nearby lay a broad *vega*, or meadow, protected from the weather by high, sheltering hills.

Henry Kimball, a blacksmith and one of Romero's friends, chose a nearby site located on the north side of the Canadian along Rita Blanca Creek, where springs poured water from the hillside. Agapito Sandoval picked a spot on Corsino Creek, sometimes corrupted to Casino Creek, on the north side of the Canadian about eight miles downstream from Romero. Eugenio Romero, a relative of Casimero, rode north, away from the others, to establish his home below the south branch of Punta de Agua, but above Minnesota Creek, at the present location of Romero. Eugenio Romero's settlement was called Romero Springs.[4]

The features of the Casimero Romero Plaza, completed in 1877, resembled those of contemporary New Mexican settlements and typified the characteristics of future *plazas* built along the Canadian. Homes with adobe walls about eighteen inches thick, corner fireplaces of adobe, and *acequias* ("irrigation ditches") to carry water from nearby springs became common. Although there was some variation, especially in the crude dugouts along Quitaque Creek, beehive-shaped corner fireplaces were standard, and most *plazas* possessed irrigation systems.

Word of abundant grasslands, clear springs, and creeks in the Canadian River Valley spread from the Casimero Romero *plaza* to New Mexico, inducing additional sheepmen to move to the free range. During the next few years much of the land along the river filled with settlers, dotting the valley with adobe *plazas*. Juan Trujillo led a group of families, including Miguel Garcia, Ysedro Sierna, and others, from Mora County. He brought the settlers to a site south of the Canadian River southwest of Cheyenne Creek. His *plaza* possessed a considerable number of sheep.

There were others. Theodore Briggs took his family to the Cana-

[4]Romero, "Spanish Sheepmen on the Canadian," pp. 47–48.

dian region from Mora County. Casimero Romero, a personal acquaintance, recommended a location that Briggs accepted. Located north of the Canadian, about one and a half miles above Rica Creek, the site stood six miles west of Casimero Romero's *plaza*. Briggs, who in 1874 had been on a military expedition into the impressive Canadian area, was returning permanently to the beautiful valley he had seen earlier.

Before 1878 a Frenchman called Padre or Father Green entered the region. Green operated a ranch above the mouth of Rita Blanca Creek about twelve miles directly south of the present town of Channing. Claiming to be a Catholic priest collecting for the church, Father Green demanded a 10 percent tithe from New Mexican settlers. Though he had little time to administer the sacraments, Padre Green proved diligent in collecting tithes, asking for 100 sheep for each 1,000 a resident possessed. Neglecting to deliver the tithes to the church, Green accumulated, as a result, a vast flock. His sheep produced so much wool that he built a large adobe storehouse on the Rita Blanca to collect the wool before taking it to market. Before 1880 he sold his ranch and livestock and left the Canadian, no doubt searching for other opportunities.[5]

Venturo Borrego, a wealthy New Mexican, during the winter of 1878–79 established a *plaza* that had about twenty-four houses. He located the settlement on the south bank of the Canadian about one mile southeast of the Casimero Romero *plaza*. Built against two small sandstone buttes, La Placita de Venturo, as it was called, stood northeast of the site of the Fort Worth and Denver railroad station at Tascosa. For a short time one of Borrego's four sons taught a small school at the *plaza*. Borrego owned several sheep and brood mares and later made a considerable profit selling horses to cattlemen.

Near the Texas–New Mexico state line stood three *plazas*. Salinas Plaza, which held perhaps twenty-five families, was located near a saline lake, from which it received its name. Its citizens used a furnace to evaporate water from the lake to obtain salt to sell. People from eastern New Mexico bought most of the salt. Not simply a salt-mining town, Salinas had stores and saloons, unlike other New Mexican settlements

[5]J. Phelps White to J. Evetts Haley, "Beginning of the LIT Ranch," January 15, 1927, Archives, Panhandle-Plains Historical Museum, Canyon, Texas; McCarty, *Maverick Town*, pp. 155–56, 178.

in the Panhandle. It became a favorite stopping place for people traveling to and from Las Vegas, New Mexico, the most substantial town in the area. Boquilla Plaza was located below Salinas Plaza. A relatively large settlement, Boquilla was populated by approximately a hundred people. East of Salinas was Chavez Plaza, possibly founded by Juan E. Chavez. The settlement, on the east bank of Chavez Arroyo at the mouth of Chavez Canyon, should not be confused with Chavez Plaza in eastern New Mexico.

About 1877 two Englishmen, Jim Campbell and A. B. Ledgard, forming a partnership, entered the Canadian. They established a ranch on Rita Blanca Creek northwest of Tascosa. The two men managed perhaps 25,000 head of sheep and 3,500 head of cattle. In the 1880s they separated, after selling their range.

An English-Scot concern, the New Zealand Sheep Company, came into the valley during the late 1870s. The company grazed thousands of sheep along Alamocitos Creek. Accounts of the number of sheep owned by the firm differ, ranging from 60,000 to 100,000 head. During the severe winter of 1880–81, freezing winds drove the company's *pastores* and their flocks southwest onto the Plains, where, unprotected from the weather, many *pastores* and a large number of sheep died from exposure or froze to death.[6]

There were dozens of other sheepmen who entered the Panhandle, some to work as *pastores* for big operators and some to establish their own concerns. They erected many ranches. Edna Kahlbau, writing in the *Amarillo Sunday News and Globe* some fifty years ago, located several settlements:

Salinas Plaza—Section 66, Block B-7, Oldham County;
Trujillo Plaza—Section 13, Block B-5, Oldham County;
Valdez Plaza—Section 111, Block 47, Oldham County;
Ortega Plaza—Section 68, Block G-M, Oldham County;
Chavez Plaza—Section 33, Block B-7, Odham County;
Tecolate Plaza—on Matador Ranch, near New Mexico;
Agripeta Sandoval—on Tierra Blanca Creek, Deaf Smith County;
Tecovas Springs Plaza—Tecovas or Sanborn Springs, Frying Pan Creek;
Juan Domingo Plaza—Section 83, Block 5, Oldham County;
Ventura Plaza—west side of Juan Domingo Plaza;

[6]John Arnot, "My Recollections of Tascosa before and after the Coming of the Law," *Panhandle-Plains Historical Review* 6 (1933): 61; Harry Ingerton to J. Evetts Haley, June 19, 1937, Archives, Panhandle-Plains Historical Museum, Canyon, Texas.

Casimero Romero Plaza—"Hogtown," Tascosa (old town) section east of original townsite;
Atascosa Plaza—Atascosa Creek, exact location of Tascosa (old town);
Joaquin Plaza—Possibly same as Tecolate (south side Canadian River, Oldham County).[7]

The kind of sheep the *pastores* grazed near their *plazas* is a matter of dispute. Most authorities maintain that the animals were a mix of Spanish *chaurro* and Merino. *Pastores* in New Mexico had for nearly three centuries grazed the gaunt animal. Lean and small, it throve on the sparse range of the upper Pecos and Rio Grande valleys. The Merino, the oldest breed of sheep in the modern world, was large, hardy, and able to walk great distances. As noted in an earlier chapter, Merino-*chaurro* crossbreeds proved to be excellent range sheep in the Southwest.[8] To the *pastores*' flocks in the Canadian River Valley sheepmen introduced thousands of Merinos. A. B. Ledgard and his partner brought several thousand Merinos to the Panhandle. They got them from Missouri, Vermont, New York, and Michigan. In 1880 their average wool clip was 4½ pounds per head for Merino-*chaurro* crosses. That compared with *chaurro* clips that averaged less than 2 pounds per head.

In grazing the animals, most, but not all, of the *pastores* used the ancient Spanish *transhumante* system of moving flocks across the country from summer range to winter range and back again. The Spanish used it in part because they believed good wool could not grow on sheep not strengthened by marching, but more importantly to keep the animals in temperatures best suited to them—highlands in the summer and lowlands in the winter. Although the practice was continued for the same reasons in New Mexico, in the Texas Panhandle the *pastores* moved their sheep simply to provide them with the best possible pastures.[9]

The Texas Panhandle flocks, following the circuits into the open range, began their journey from the Canadian River Valley early in the

[7] Edna Kahlbau, "Ghostly *Plazas* Once Rang with Songs of Shepherds," *Amarillo Sunday News-Globe*, Golden Anniversary Edition, August 14, 1938, Section C, p. 4.

[8] Edward N. Wentworth, "Sheep Trails of Early Texas," *Southwestern Sheep and Goat Raiser*, pt. 1, 9 (June, 1939): 7.

[9] For a detailed discussion see Julius Klein, *The Mesta: A Study in Spanish Economic History*, pp. 1–220.

year to ensure arrival on the plains' pastures by lambing time. The *pastores* led the flocks through the open range during the summer months, circling back during the later months in order to return to their home range by shearing time. Some owners chose to keep the sheep on the range and sent groups of workers and large wagons to meet the flocks fifty or sixty miles out from the settlements to shear the animals and carry the wool to market. The circuits, which tended to concentrate where suitable sheep conditions existed, followed closely the river valleys, creek beds, and canyons, where abundant grass and sufficient water could be found. Such locations were scarce in the Panhandle, except after spring rains, when sufficient water could be found in the large circular depressions characteristic of the High Plains. The northeastern region, nearly one-fifth of the Panhandle, was infested with locoweed, making it unfit for raising sheep. About three-eighths of the southwestern region was part of the Llano Estacado and lacking in water. Since raising livestock required a close location of the precious resource, especially during the hot summer months, only a few herders entered the Llano Estacado. The areas that the sheepmen favored were also preferred by cattlemen, a factor that later caused discontent between the two factions.[10]

Grazing circuits extended into the Red River drainage system as far southeast as Tule and Quitaque canyons and beyond into the Brazos River basin. Jesús Perea herded his flocks of thirty thousand sheep to Tahoka Lake, Yellowhouse Canyon, and Blanco Canyon. Because of the great amount of grazing land and water needed for so many sheep, Perea scattered his animals widely and took them wherever good grass and water could be found on the South Plains.[11]

As the area filled in with ranches, sheep-grazing circuits became more and more fixed. As a result, some *pastores* built rough stone corrals on the open range to shelter themselves and their sheep. Some used other forms of protection. Casimero Romero used wagons to transport lumber panels, which served as portable sheep pens for his flocks. When they returned from the range to the settlement for the winter, the sheep usually bedded inside stone or adobe corrals con-

[10]T. R. Havens, "Sheepmen-Cattlemen Antagonism on the Texas Frontier," *West Texas Historical Association Year Book* 18 (1942): 10–23.

[11]Romero, "Spanish Sheepmen on the Canadian," p. 47.

nected to, or close by, the *plazas*. Occasionally an owner chose a grassy spot with a good water supply near a hill or bluff and built a fence to enclose the area.[12]

Men with large flocks, like Ledgard and Perea, who grazed several thousand head in the Panhandle, adopted the Spanish system for dividing work among four types of employees: the lowest in rank was the *pastor*, who kept watch over a flock of about 1,500 sheep. Over two or three *pastores* was a *vaquero*, who selected the watering places and the area to be grazed for the day. A *caporal* had charge of several *vaqueros*, and the *mayordomo* supervised the *caporales* and accepted responsibility for the entire operation. Wages for Mexican herders were reported to be $15.00 monthly, and board averaged about $4.75 per month. Many good Indian herders, including Pueblos and Navajos, entered the Panhandle from New Mexico. In the Panhandle sheepmen with large operations maintained the flocks in bands of 2,500 to 3,000 head, with two or three herders to a band. Along the Canadian, however, most sheep were raised in small flocks for which the owner did most of the work himself, including the lambing and shearing.

At the western, or upper, end of the Canadian the *partido* system prevailed after about 1880. Although used elsewhere in Texas and in New Mexico for half a century or more, its introduction in the Panhandle came late. New Mexican merchants and *ricos*, as the large ranchers were called, saw little profit in sheep husbandry until the national economic depression of the mid-1870s and the spread of railroads brought keener local and intersectional marketing competition. The *partido* system, if used on a selective basis and carefully supervised to prevent abuses on the part of the *partidarios*, was a sound method of refinancing or funding debts of customers whose account balances with Las Vegas merchants had grown unduly or had not been reduced. Thus, many merchants placed ewes on shares with selected farm and ranch customers. A common practice found merchants and customers sharing the wool clip and dividing the lamb crop. The herder could sell his surplus animals to the merchant to reduce his debt, retain the original sheep on a rental contract, and enjoy extra profits from the wool. Over an extended period, the *partidario* eliminated the debt and built up a sizeable flock.[13]

[12] Ibid., pp. 47, 60.

[13] William J. Parish, *The Charles Ilfeld Company*, pp. 150–60.

Partidarios marketed their sheep and wool in Las Vegas, New Mexico. Most of the other *pastores* may have sold there, too—for example, Casimero Romero hauled his wool to the old mountain city. Some, though, transported their sheep and wool to Dodge City, Kansas, the nearest railway shipping point. O. H. Nelson, an early Panhandle cattleman, reported that in a single year one family drove twenty thousand sheep to Dodge City and that Mariano Montoya, who in 1878 had settled along the confluence of Rita Blanca Creek and Punta de Agua, likewise took sheep to the Kansas rail center.[14] Antonio Baca, the Oklahoma Panhandle sheepherder, marketed his wool and wethers in Kansas. Some men trailed their wethers to Santa Fe, New Mexico, or to Trinidad and Pueblo, Colorado, or sold them to eastern feeder operators for fattening.

As the *pastores* prospered, cattlemen challenged the sheepmen for the land. The first to arrive was Charles Goodnight. A former cattle trailer who had opened a route from Texas to Colorado via New Mexico, Goodnight, suffering some business reverses in Colorado, moved his herd of 1,500 cattle southeastward into the Panhandle. After wintering among the sheepmen, in 1876 he entered Palo Duro Canyon of the Red River. There, in partnership with John Adair, an Irish capitalist, he established the JA Ranch. Goodnight agreed to stay away from sheepmen.

With other cattlemen the *pastores* were not so lucky. In 1876 Elsworth Torry, an old Yankee sea captain, laid out a ranch with 4,500 cattle on the south bank of the Canadian near Skunk Arroyo, the first creek east of the Alamocito. The next year, Thomas S. Bugbee established the Quarter Circle T Ranch in Bugbee Canyon, formed by a tributary of the Canadian, only seventy-five miles below Tascosa. The same year organizers of the LX Ranch, W. H. Bates and David T. Beal, moved several thousand cattle onto the Canadian sheep range. George W. Littlefield brought 3,500 head to the mouth of Pescado Creek and soon bought Henry Kimball's homesite for his ranch headquarters.

By 1880 the Panhandle contained a sizeable population, totaling more than 1,500 persons, including nearly 400 *pastores* and other sheepmen, with perhaps 108,000 sheep. There were more than 97,000

[14]O. H. Nelson to J. Evetts Haley, February 26, 1927, Archives, Panhandle-Plains Historical Museum, Canyon, Texas.

cattle feeding in the Panhandle.[15] The animals ranged from below Yellowhouse Canyon on the south to the Oklahoma-Texas border on the north. After 1880 the number of sheep in the region declined drastically, until by 1890 there were about 10,000 head, nearly all of which were grazed by cattlemen.[16]

Before the decline, however, the *pastores* in their *plazas* developed an unhurried and uncomplicated existence. Almost anything they wanted to eat could be had from the land. They made occasional trips to New Mexico trading points, especially Las Vegas, for other items, but their diet and clothing remained simple. Fishing, hunting, trapping, cock fights, and rooster races took up much of the time away from the flocks. Birds for hunting, especially turkeys, were plentiful. Wild game included wolves, coyotes, antelope, elk, buffalo, and others. Wild fruit was abundant. Games also were prevalent. The favorite perhaps was *La Pelota*, which resembled field hockey. The participants played the game on a clearing longer than a modern football gridiron with equipment made from native materials. The New Mexican sheepmen also enjoyed fiestas. Well-attended by the valley families, the fiestas were lively affairs at which dancing, singing, eating, and drinking often continued through the night. Cowboys from the cattle spreads rode their horses many miles to attend a *baile*, or dance, eagerly anticipating the chance to waltz with *señoritas* and to drink the *pastores*' liquors.[17]

Although sheep raising was profitable and life was simple for the *pastores*, the occupation was not without its difficulties. The often repeated tale of the John Casner family and Sostenes Archiveque caused many sheepmen in 1877 to abandon the Canadian Valley almost immediately after they had arrived. In the mid-1870s John Casner and his three sons decided to leave California, where they had made a fortune in the gold fields, to raise sheep in the Texas Panhandle. Two of the

[15] Seymour V. Connor, "Early Ranching Operations in the Panhandle: A Report on the Agricultural Schedules of the 1880 Census," *Panhandle-Plains Historical Review* 27 (1954): 49–69; Ernest R. Archambeau, "The First Federal Census in the Panhandle, 1880," *Panhandle-Plains Historical Review* 23 (1950): 25–27.

[16] "Sheep and Wool on Farms by Counties," in U.S. Bureau of the Census, *Eleventh Census. Report of the Statistics of Agriculture: 1890*, pp. 268–69; Rathjen, *Texas Panhandle Frontier*, p. 244.

[17] Pauline Durrett Robertson and R. L. Robertson, *Panhandle Pilgrimage*, pp. 134–35.

sons started for Texas with an ox-drawn wagon, a few head of horses and cattle, and a flock of sheep herded by a Navajo boy. They planned to rejoin their father and brother, who had taken another route, on the Palo Duro range. Unfortunately, Sostenes Archiveque, a ruthless outlaw, camped along the Canadian River at the time the Casner brothers entered the Panhandle. Wanting the Casners' money, perhaps twenty thousand dollars in gold coin, Archiveque ambushed and killed the brothers; the gold disappeared. For committing the crime, *pastores* in the valley stabbed Archiveque to death, but John Casner and his other son, Lew, when they arrived, were not satisfied. They determined to avenge the deaths of their kin. A reign of terror followed in which the Casners killed or threatened several of the New Mexican sheepmen and their families. Mexican messengers, riding day and night to all the *plazas* along the Canadian, spread the news of "those Californians" and their terrible revenge for the murder of the Casner brothers. Families deserted several *plazas* and sheep camps, returning to their former homes in New Mexico to escape the Casners, who finally took their sheep, which had been cared for by Goodnight's cowboys, and went to New Mexico.[18]

Most of the *plaza* dwellers remained, however, riding out the trouble. But shortly afterward they encountered other difficulties. During the early 1880s a series of winters with freezing, roaring blizzards brought considerable losses to the sheepmen. Moreover, cattlemen purchased title to much of the land that *pastores* had claimed for years, thus reducing the open range sheep industry. After gaining title to the land, the ranchers drove out the *pastores* and tore down or burned their homes. When *pastores* did not leave quietly, the cattlemen drove them off at gunpoint, set fire to the sheep range, or drove the sheep over small cliffs.

Between 1874 and 1883 the Texas legislature passed a series of laws that affected the entire Texas sheep industry, but especially the Panhandle *pastores*. The first of these laws, passed in May, 1874, concerned the migratory sheepherder, or drifter. The drifter was disliked by settled sheepmen as well as by the politically powerful cattlemen,

[18] Ibid., pp. 138–39; McCarty, *Maverick Town*, p. 34; J. Evetts Haley, *Charles Goodnight: Cowman and Plainsman*, pp. 280–90; J. Evetts Haley, "Pastores del Palo Duro," *Southwest Review* 19 (April, 1934): 279–94.

who lobbied for the law in Austin. In the Panhandle drifters moved their flocks in from New Mexico in the summer and back again in the fall. The law made the grazing of infected sheep on land not belonging to the sheepman or the driving of such sheep on a public road a misdemeanor. It was designed to keep migratory flocks, often infected with communicable diseases, out of the open range. It slowed, but did not stop, the migration of drifters and *pastores* to the Texas Panhandle.[19]

A second law, passed in March, 1881, also dealt with drifters. Under this law, county inspectors were appointed to examine all sheep present in their county, including flocks belonging to drifters. If an inspector found a flock infected with scab, a highly infectious disease that could contaminate the soil, he was to notify the owner or herder and to restrict the flock to a certain area. Although this law discouraged some New Mexican herders, by hiding diseased sheep before inspection, drifters often obtained clean bills of health for their animals. If the diseased sheep were found, drifters simply disclaimed ownership. But when tightly enforced, the law proved effective. It reduced the number of drifters and consequently *pastores* in the Panhandle.[20]

A month later the Texas legislature passed another law that restricted Panhandle *pastores*. This law forbade the driving of sheep onto land not belonging to the herder when the landowner objected. If he did not move the flock, the sheep owner would be liable for a fine of up to one hundred dollars. When the law proved ineffective, an amendment was added in 1883, which provided that sheepherders must carry a certificate of inspection issued less than sixty days from the date the sheep were moved. Since most of the *pastores* had no legal title to the land they claimed, the new law not only stopped the influx of additional *pastores* but restricted those already on the land.[21]

Taxes also cut into the ranks of the *pastores*. When the region was thinly populated, drifters moved their sheep into Texas before New Mexico tax collectors could assess duties on the animals and back to New Mexico in time to evade Texas tax men. In the 1880s, however, the local sheriff at Tascosa began collecting a drift tax on both sheep

[19] T. R. Havens, "Livestock and Texas Law," *West Texas Historical Association Year Book* 36 (1960): 28.

[20] Havens, "Sheepmen-Cattlemen Antagonism," pp. 20–21.

[21] Ibid., pp. 21–22, 32.

Mexicans and carts, drawing by Frank Arthur Stanush. Courtesy, Texas Sheep and Goat Raisers Association and Hiram Phillips.

Mexican sheep shearers, near Fort McKavett. From U.S. Department of Agriculture, *Special Report . . . Condition of the Sheep Industry*, 1892.

Shepherd and flock, on George Richardson ranch near San Angelo, early 1900s. Courtesy, Fort Concho Museum, San Angelo.

Prototype of two-wheeled Mexican *carreta* ("cart") used to haul wool, especially from South Texas to Corpus Christi. Courtesy, Mission San José y San Miguel de Aguayo, San Antonio.

Shearing sheep on Case ranch near Eldorado. Courtesy, Fort Concho Museum, San Angelo.

Mexican sheepherders and shearers, 1890s. Courtesy, Fort Concho Museum, San Angelo.

Left: George Wilkins Kendall. From Fayette Copeland, *Kendall of the Picayune*, following page 308. *Right:* William L. Black. From *Sheep and Goat Raisers* 30 (July, 1950): 4.

Left: Sam Hill. Courtesy, Southwest Collection, Texas Tech University. *Right:* Charles Schreiner. Courtesy, Southwest Collection, Texas Tech University.

and cattle herds moving into the Panhandle. The number of out-of-state livestock owners soon declined. Drifters with sheep disappeared.[22]

But it was barbed-wire fencing that reduced the open sheep range and sealed the fate of the Panhandle *pastores*. Joseph F. Glidden, a farmer of De Kalb, Illinois, had invented a superior barbed wire by 1874. Henry B. Sanborn, a salesman for Glidden wire and later one of the founders of Amarillo, at first found it difficult to sell his product in Texas. But by 1880 there were a few barbed-wire fences in a number of counties. In the Panhandle, the T-Anchor ranch was the first large cattle spread to enclose its range. In 1881–82 it fenced 240,000 acres of grassland. In 1883, Glidden and Sanborn fenced, at a cost of $39,000, the 250,000-acre Frying Pan Ranch. Goodnight and other cattlemen quickly followed these leaders, reducing even more the land open to sheepmen from New Mexico.[23]

Some people bought out the *pastores*. W. M. D. Lee, founder of the LE and LS ranches, who had grown wealthy in the freighting business at Dodge City, bought out many sheepmen. One morning in 1882 he started out with a bag full of money at Trujillo, the westernmost *plaza*. Riding down the Canadian in his buggy, Lee lost little time at each settlement, providing cash from his valise and demanding that the *pastores* get out in return for the money. His method was effective; "it wasn't long after a visit from Lee that most [sheepmen] began leisurely grazing their flocks . . . back toward New Mexico." The herders loaded their women, children, and household goods on wagons to complete the exodus.[24]

By 1884 most of the *plazas* were deserted. Some became line camps or station houses for cattlemen. Others were razed or burned. Before many years passed the sod roofs of those not destroyed sifted down to the floor. Rabbits and ground squirrels burrowed under the foundations, and "shifting sands of streams, long years in sun and wind . . . wrapped the *plazas* in silence and ruin."[25] All that is left of most of

[22] McCarty, *Maverick Town*, pp. 165–66; White to Haley, "Beginnings of LIT Ranch."

[23] Haley, *Charles Goodnight*, pp. 222–23, 321; Taylor, "*Pastores* in the Texas Panhandle," p. 39.

[24] Robertson and Robertson, *Panhandle Pilgrimage*, p. 140.

[25] Kahlbau, "Ghostly *Plazas*," p. 4.

them is their Spanish names, and, in a few cases, dilapidated stone corrals or decaying rock foundations.

The sheep outfits that remained in the Canadian River Valley were those which combined sheep with cattle. One *pastor*, Antonio Trujillo, with his large family, raised both sheep and cattle until a cattleman claimed that Trujillo's sheep had infected his horses with mange. After the charge, Trujillo abandoned his sheep enterprises. Casimero Romero in 1882 changed from livestock raising to freighting, running a business between Tascosa and Dodge City. In 1896 Romero finally returned to New Mexico. Agapito Sandoval, the last of the New Mexican *pastores*, remained at his *plaza* on Corsino Creek until 1887.

Although their stay in the Panhandle was brief, the *pastores*' part in the Texas range sheep and goat industry was large. *Pastores* opened the Panhandle. They located waterholes, built homes, and encouraged a settled life on the empty and isolated range. They developed trade routes that *comancheros* and buffalo hunters had used between the Panhandle and Las Vegas and Dodge City. They showed that sheep could graze successfully in the region and demonstrated—though not to the cowboys' immediate satisfaction—that sheep and cattle could share the same ground. As the *pastores* quietly vacated the Panhandle, sheep and goat raisers stampeded onto the vast and semiarid Edwards Plateau, one of the finest sheep- and goat-producing regions in the world.

7
The Edwards Plateau

THE Edwards Plateau, comprising all or portions of thirty-seven counties and embracing some twenty-two million acres, is the chief area of sheep and goat production in present-day Texas. The region lies south of the Colorado River and west and north of the long curve of the Balcones Escarpment, which extends from the Colorado at Austin to the Rio Grande at Del Rio. Usually the Pecos River is placed as the western boundary, but Pecos and Terrell counties are sometimes included. The broken eastern part of the plateau, known as the Hill Country, is covered with cedar on the uplands and great growths of pecan and other hardwoods along the streams. An extension of the Great Plains, the western plateau is a ranging tableland.

Most of the Edwards Plateau is covered with curly mesquite grass but has a sprinkling of buffalo, grama, needle, and bunch grasses. There are many brushy thickets of live oak, shin oak, mesquite, cedar, sumac, and cat claw. There are many drought-resistant plants, such as sacahuiste, sotol, yucca, and prickly pear. The elevation of the plateau varies from 1,000 to 3,000 feet. Average rainfall ranges from thirty-five inches on the extreme eastern boundary to twenty inches on the west. The summers are hot, but because of welcomed breezes, not oppressive. The average July temperature is eighty-three degrees (F.). The winters are mild, but occasionally bitter cold causes loss of livestock. The average January temperature is forty-four degrees. The region is drained by several rivers, including the Devils, Nueces, Llano, Guadalupe, San Saba, Concho, and Colorado. The soils are thin and of limestone origin, but considerable farming is done in Concho, McCulloch, Schleicher, Sutton, and Tom Green counties. Cotton, grain sorghums, and oats are the main cultivated crops. The region, however, is principally one of livestock raising. San Antonio on the southeast, Austin on the east, and San Angelo on the north serve as distributing and banking centers. San Angelo, Del Rio, Fredericksburg, and Kerrville are the main concentration and shipping points.[1]

Before 1875 sheepmen who entered the unsettled western area of the Edwards Plateau faced the threat of Indian attacks. During the rainy seasons of the spring and fall, Comanches, Apaches, and others hunted in the plateau, stealing sheep when they found them. Thefts of livestock were frequent. But after 1875 Indian raids declined and then eventually stopped. As soldiers drove Indians out or penned them up on reservations, stockmen filtered into the plateau. The first herders stuck close to watersheds. With the development of wells, range tanks, and other watering equipment, grazing spread out over the isolated and semiarid country. The Hill Country filled in first, long before 1875. Then sheepmen followed the rivers westward. The territory west of the 100th meridian to the Pecos River was, as far as wool production was concerned, in its infancy. About 1875 a western line of farms extended from Clay County on the north through Archer, Young, Stephens, Comanche, Mills, San Saba, and Mason counties. West and south of that line was sheep, goat, and cattle country.

Some of the early sheepmen in the region beyond the Hill Country included John Mundry, Joseph Tweedy, Morgan Grinnell, J. Barlow Reynolds, and Captain William Turner in Tom Green County, John Arden in Concho County, William L. Black and Jim Shannon in Schleicher and Menard counties, John Rae in western Schleicher County, P. H. Wentworth in Sutton County, and Ainslie Turner in Sterling County. There were many others, of course, especially in the period after 1880, when the Texas sheep boom exploded over the Edwards Plateau. These daring adventurers faced potential threats from several areas in the late 1870s. Since almost without exception they appeared after cattlemen had established themselves, wool growers were seen as unwelcomed intruders. In addition, wolves and coyotes attacked isolated sheep. A few Indians from Mexico, wandering through ancient hunting land, drove off sheep until after 1879. Buffalo hunters still ranged over the South Plains, disrupting flocks as they butchered

[1]See *Texas Almanac*, 1980–81, pp. 100–102, 112–14; Harold M. Gober, "The Sheep Industry in Sterling County," *West Texas Historical Association Year Book* 27 (1951): 32–33.

the last of the mighty brown beasts for their hides. In 1877 they shipped 100,000 hides out of San Angelo.

Sheepmen on the Edwards Plateau marketed the first wools to T. C. Frost of San Antonio. Later they sold their product through Charles Schreiner at Kerrrville, Theodore Heick at Abilene, and Charles W. Hobbs and J. H. Meara in San Angelo. Wethers they often kept to increase the size of flocks, or they sold them locally. The marketing of lambs began about 1900. Sol Mayer, who arrived at Fort McKavett in 1879, claimed he loaded the first shipment of lambs from San Angelo in 1899. They sold in Kansas City. "I do not think," he afterward wrote to Edward Wentworth, "very many lambs were shipped out for several years later."[2]

The first sheep were Spanish *chaurros* bred to Merinos and Delaines, which were the American strain of Merinos. Later wool growers brought in Rambouillets, the French Merino strain. Rambouillets, which eventually displaced Merinos as the favorite improved stock in Texas, are the largest and strongest of the fine-wool sheep. Rams weigh 200 to 250 pounds, ewes 140 to 200 pounds. Compared with the Merino, the Rambouillet's body is smoother and its mutton value is somewhat greater. As a rule, Rambouillet wool is slightly coarser than Merino, but the length of the fibers is more uniform. The Rambouillet takes its name from the village of Rambouillet, France, where the royal farms were located in the days of Louis XV. Louis, desiring to do something to improve the quality of the French sheep, convinced his cousin, Charles III, the king of Spain, that he should be permitted to bring from Spain to France a band of sheep to start a foundation flock on the royal farms at Rambouillet. Rambouillets came to America in the Napoleonic era, but not until the 1880s and 1890s did they enjoy wide interest among Texas sheepmen.[3]

The two men largely responsible for the introduction of the Rambouillet breed to the Edwards Plateau were Arthur G. Anderson and Karl Albert Anton von Schauer. Anderson, born in Tennessee, came to Texas as a youth with his family in 1859 and settled temporarily in Menard County west of the present-day town of Menard, where his father

[2] Edward N. Wentworth, "The Golden Fleece in Texas," *Sheep and Goat Raiser* 24 (December, 1943): 30.

[3] See E. H. Patterson, "Romance of the Rambouillet," *Sheep and Goat Raiser* 27 (October, 1946): 10–11.

planned to establish a ranch. But the Civil War caused removal of the family, first to San Antonio and then to New Orleans, where Anderson received his education while his father served as a financial officer for the Confederacy in Europe. Frail as a child, Anderson was small of stature, quiet and shy by nature. He never married.

In 1872 he was back in West Texas, where in 1875 he established a sheep and cattle operation, ranching in Callahan and Taylor counties. Deciding later that he was not satisfied with the breed of sheep he was running, he sold out and began a search for the type he considered best suited to the Edwards Plateau. His search took him to Ohio, Vermont, and California before he settled on the Rambouillet as the type closest to what he wanted. In California he selected and bought a flock of sheep. He trailed them back to Texas along a circuitous route through Salt Lake City, Utah, and Colorado, taking two years to make the trip. He established his sheep on a 125-section spread in Mitchell County near Colorado City. They became the foundation for his future flock of Rambouillets, which he built up in after years, running at times twenty to forty thousand head. Later he established a 250-section ranch—the Hat A—in Pecos County. It was through his sale of purebred rams that many fine-wool growers got their start. He sold annually about twelve hundred animals.[4]

Karl Albert Anton von Schauer, born in Verona, Italy, in 1831, was the son of a field marshal in the Austrian Army. He came to Texas at the close of the Mexican War, and as a young man served in the United States cavalry. While stationed at Fort Sumner, New Mexico, in the late 1850s, he conceived the idea of driving purebred sheep from California to the Texas frontier. In 1859 he traveled to California, bought 3,000 head, and began three years of trailing them to Texas. He located a headquarters ranch at Ozona but grazed his sheep through Crockett and Tom Green counties. Von Schauer hired one herder for each 1,500 sheep and moved the flocks twice a month to conserve grazing. Eventually he ran about 30,000 head, most of them Rambouillets. When his

[4]"Blazing the Way," *Sheep and Goat Raisers' Magazine* 7 (December, 1926): 4; Arthur G. Harral, "Arthur G. Anderson, Pioneer Sheep Breeder of West Texas," *Southwestern Sheep and Goat Raiser* 6 (November, 1935): 12–13, 31; newspaper clipping, Morgue Files, Texas Sheep and Goat Raisers Association, Records, Southwest Collection, Texas Tech University, Lubbock.

purebreeds started attracting attention in the Edwards Plateau, he sold thousands of head to area ranchmen.[5]

Like Anderson and Von Schauer, many new wool growers on the plateau were immigrants from Europe or states of the Old South. An estimated 400,000 people entered Texas in 1876, and many of these went directly to the frontier. Menard County attracted some of them. Slightly less than 50 percent of the population in the 1870s and 1880s were German. The area was utilized almost exclusively for sheep ranching. One of the most important German ranching families in the county was the Wilhelms. John and Johanna Wilhelm pioneered there in 1882 after trying agricultural pursuits in other parts of Texas for more than a decade. When John died in 1890, Johanna with her seven children continued to operate the twenty-section ranch, on which she grazed several thousand sheep. In 1900 she owned over one hundred sections and ran between ten and fifteen thousand sheep, goats, and cattle, marketing some five thousand sheep each year. Newspapers dubbed her the "sheep queen of West Texas."[6]

Captain Herman Steiler, another figure of German descent, attained prominence in sheep husbandry. Born in Germany in 1853, he came to Texas three years later. In 1863 his family settled in Kendall County, where he got his early sheep training from Charles Schreiner. During Indian troubles following the Civil War, he served as captain of a volunteer defense organization in his community. In the late 1870s he claimed three sections of land in the Hill Country near Comfort and started raising sheep. His business prospered. He bought more sheep, invested in Angora goats, and extended his landholdings to twenty-five sections. He preferred Delaines to Rambouillets, however, as better suited for his hilly range country.[7] When he retired, his four sons continued the sheep and goat ranch. One of them, Adolf, who concentrated on Angoras, became known in the twentieth century as the "goat king" of Texas.

John Scharbauer was still another successful wool grower of German descent. Born in Albany, New York, in 1852 to immigrant parents, Scharbauer moved to Texas in 1880, engaging in sheep raising. As time

[5]Wentworth, "Golden Fleece in Texas," p. 31.

[6]Cited in *History of the Cattlemen of Texas*, p. 247.

[7]Edward N. Wentworth, *America's Sheep Trails*, p. 386.

passed, he moved farther west, finally locating near Midland in 1887. The following year he began raising cattle and expanding his ranchland. In 1895 he owned all or parts of at least six ranches totaling about 200,000 acres. He opened a bank, but was known primarily for breeding high-quality cattle. From his ranch property, scattered over a large portion of West Texas, he marketed as many as 47,000 head of sheep in a single year.[8]

Germans were not the only immigrants on the Edwards Plateau. In 1876 Ernest Carlin, a Frenchman, established his headquarters ranch about six miles southwest of the present town of Menard. He proceeded to acquire some 200,000 acres of good sheep land in the region. Although it is not known where he got his money, it was commonly accepted that he came by it illegally. Carlin built a lavish rock house on a hill where it commanded a view of, and could be seen from, many miles in every direction. A sixteen-room mansion with a cellar stocked with fine French wines, dried fruits, and canned delicacies, the residence was the scene of expensive entertaining. Guests, arriving regularly from all over the state, enjoyed the finest hospitality Carlin could provide.

But Carlin, spending too much time on his guests, mismanaged the ranch. The lavish expenditures for entertainment broke him. Others assumed control and adopted the name Los Moras Ranch Company. When Carlin returned to Paris, H. C. North became the foreman, operating the spread until Carlin's death in 1887, at which time the property was divided. Louis Runge took charge of the Los Moras portion, consisting of 115,000 acres. Within a few years he had imported blooded stock, set up windmills, fenced it into separate pastures, and generally improved the property. The Los Moras–Runge spread ran more than 9,000 cattle, 10,000 sheep, and 2,000 goats. Los Moras became one of the most prosperous ranches in the state.[9]

In many ways Britain provided the most typical of the early sheepmen. Scotsmen, looking for a quick and sure profit, and Englishmen, unwilling to acknowledge any personal slight associated with working

[8] James Cox, *Historical and Biographical Record of the Cattle Industry and Cattlemen of Texas and Adjacent Territory*, p. 432.

[9] "French File," Institute of Texan Cultures, San Antonio. See also Bobby Weaver, "German Contributions to the Texas Cattle Industry," (1977 typescript, photocopy in Southwest Collection, Texas Tech, pp. 19–20.

sheep, spread through the Hill Country and beyond. Thomas Hughes, Jim Shannon, William Adams, G. F. Ruxton, R. B. Townsend, John Clay, William French, W. S. Shepherd, Thomas Carson, Baille Graham, Charles Gordon, Harry and Robert Maudslay, W. G. Sutherland, C. P. Campbell, Robert Boyle, Andy Mansfield, John Bennett, and Thomas and Harry Holdsworth represent only a few. The Britishers were usually well educated and often wrote of their experiences. Thomas Hughes, who settled near Boerne, took time off from his sheepherding to put together a book entitled *G. T. T.—Gone to Texas: Letters from our Boys*. In the book, published in 1884, Hughes sought to inform other Englishmen on matters of Texas sheep. Robert Maudslay produced a valuable journal that his niece Winifred Kupper used as the basis of a lively tale about sheep husbandry in the Southwest. John Clay wrote *My Life on the Range*. Ruxton, Townsend, French, Shepherd, Carson, and Graham also recounted their impressions and experiences in the West. Their works add substantial information to the history of the Texas range sheep and goat industry.[10]

Americans also pressed into the Southwest. Four New Yorkers, whose story resembles those of many early wool growers, established one of the earliest sheep operations south of San Angelo and helped to found the town of Knickerbocker. In the spring of 1876 Joseph Tweedy, Morgan and Lawrence Grinnell, and J. Barlow Reynolds, on the wave that carried thousands of young men to Texas, sailed for Galveston with modest capital and several books on sheep. Their plan was simple: to learn all details of sheep husbandry by practical experience and to become very rich. Their success was mixed. While only Tweedy prospered in the sheep industry, they all learned the business. From Galveston they went to Fort Clark at Brackettville, bought *chaurro* ewes and Merino bucks, and settled down to watch their sheep and study the business.

In the brush- and chaparral-covered country between Fort Clark and Eagle Pass the partners camped for a year. Here along the breaks of a small river they shared the colorful range with a pair of greenhorn easterners like themselves. Dipping for scab, nursing an unexpected lamb crop, and spending days and weeks with the flock through bleak

[10] Florence Fenley, *Old Timers, Their Own Stories*; Thomas Hughes, ed., *G.T.T.—Gone to Texas: Letters from our Boys*; Winifred Kupper, *The Golden Hoof: The Story of the Sheep of the Southwest*; John Clay, *My Life on the Range*.

northers, drenching rains, and oppressive heat, they learned the southwestern sheep industry. Indian raiders forded the Rio Grande on moonlit nights to steal livestock. Stray Englishmen appeared through the chaparral. Coyotes, wolves, and other varmints surrounded their flock. When cattle drifted into their territory, they discovered that sheepmen did not tell cattlemen to keep their cows away. No problem resulted, for the partners discovered that cows ate the tall grass that sheep did not want and left enough underneath for the sheep to get at.

In their camp they ate simple meals of coffee, beans, bacon, jerked meat, and corn dodger, occasionally supplemented by beef, a can of tomatoes, an onion, or a rabbit. They developed scurvy from lack of fruit. Lawrence Grinnell never fully recovered from the ailment and later returned to New York, where he died in 1881. They discovered the diverse uses of mesquite—as fuel, as pins for mending clothing, as shade for sheep and men. They learned to sleep on hard ground, under wet blankets, and through cold winter rains. One night when their sheep suddenly drifted off they found that good herders sleep in their clothes. They also developed the habit of sleeping with one ear tuned to the habits of the flock. In the evenings around the campfire or a lantern in the tent, the four young men read through boxes of books and wrote long letters home. That year the country celebrated its centennial, and they, despite being in an isolated and far-off corner of the nation, did not plan to be left out. On July 4 they had their own Independence Day party. The year was also a presidential election year—Samuel Tilden running against Rutherford B. Hayes. Around the campfire one evening in the fall they held a lone Republican rally for Hayes.

They developed a distrust for Mexicans. In part it stemmed from the general feeling against Hispanics in that area of the country. But it was not all ethnic animosity. Across the Rio Grande there were a few Mexican ranchers who hired gunmen to enter Texas to steal sheep and cattle to stock their own ranges. Depredations from these outlaws were heavy enough to make every Mexican suspect. Anglos like these partners made little effort to hide their feelings for the "greasers," as the Mexicans were called, and more than one Mexican retaliated with robbery or a worse crime. Although West Texas had a plentiful supply of outlaws at the time, only a tiny percentage were of Mexican descent. But the Mexicans, according to at least one scholar, were quicker than others to take affront, and the Anglos knew little about how to get along with them. A cultural clash developed.[11]

When they moved northward in the spring to the country the four men had claimed, south and west of Fort Concho at San Angelo, the New Yorkers' train had increased. They had acquired over the year, besides additional stock and experience, two *pastores*, two wagons and teams, a dog, and a young Englishman who had joined the outfit as a cook. He was "chiefly noted," they wrote, "for being able to parch coffee and read Shakespeare at the same time." By slow stages, making about seven to ten miles a day, they moved up the valley of the west fork of the Nueces, across the divide to the head of the South Llano River, and thence to Fort McKavett. After resting briefly here, they moved by way of the Eight-Mile Draw to the Concho and on to a camp on Dove Creek in present Tom Green County. They named the ranch Knickerbocker after a favorite character in Washington Irving's writings.[12]

The first Knickerbocker headquarters was a thatched adobe shack papered with copies of *Harper's Weekly* and *Frank Leslie Illustrated*. The partners built pens, a brush shelter, and a ranch store. Two miles down Dove Creek a community of farmers was already settled, with a church and a schoolhouse in full operation. When Morgan Grinnell, Tweedy, and Reynolds in 1881 applied for, and got, a post office, the two settlements merged into the small but flourishing town of Knickerbocker. Because most good land along creeks had been taken by private patents, many of them in the names of the earliest German settlers, the partners leased and later bought (at seventy-five cents per acre) land along the larger streams and rivers. Back from the water they spread their sheep over wide areas, invading railroad lands and other property. In 1884 a San Angelo newspaper reported that their range covered 100,000 acres in a solid body about twenty miles long and eight miles wide, extending from Spring Creek to the South Concho. The partners established sheep camps at Burkes Creek, on Spring Creek, and near the head of Dove Creek. As their operation expanded, they brought in additional *pastores*, who worked for twelve dollars a month plus board, *caporales*, and *vaqueros*.

[11] See, for example, Kupper, *The Golden Hoof*, p. 82.

[12] *San Angelo Standard-Times*, clipping, Morgue Files, Texas Sheep and Goat Raisers Association, Records, Southwest Collection, Texas Tech.

Grinnell, Reynolds, and Tweedy faced the usual problems. Flocks disappeared in snowstorms. Prairie fires that swept the range drove sheep away before them. Scab infested their flocks. Coyotes picked out the stragglers. Mother ewes refused to suckle their young. "If we stick to it," wrote Joseph Tweedy to his family in New York, "we are bound to make money. If we did not think so, I don't believe any of us would stay in Texas a day longer than necessary."[13] They persevered. Indeed, they watched their small band grow large. In 1881 they sheared 51,000 pounds of wool, which they sold for twenty-five cents per pound to a Boston firm, shipping the product east through Abilene, then a thriving little town only a few months old. Sharing the more than twelve thousand dollars in profits from the sale, the partners seemed well on the road to success. They hired more herders, bought more land, and improved their ranch. Conditions were so good that Tweedy and Grinnell went to New York briefly to marry before returning with their brides and wagons loaded with household goods.

But two years later the bottom dropped out of the wool market. Prices plummeted. As the value of sheep and grazing land subsequently dropped to almost nothing, overextended sheep ranchers went down in great numbers, among them Grinnell, Tweedy, and Reynolds. Grinnell and Reynolds withdrew. Joseph Tweedy stayed, continuing to operate on the Dove Creek lands he was able to save when the partnership collapsed.[14]

By this time the Texas sheep boom was in full swing. It attracted thousands to the Edwards Plateau. They came to raise sheep, to farm, or to operate a business in the scores of towns that sprang up, like Knickerbocker, to serve the farmers and ranchers. Railroads, extended westward to the plateau and around its edges, facilitated the migration. The Corpus Christi, San Diego, and Rio Grande line reached Laredo in 1881. Early in 1882 the Texas and Pacific joined the Southern Pacific at Sierra Blanca in far West Texas. The San Antonio and Aransas Pass Railroad expanded as far as Kerrville in 1887, and the Santa Fe was built westward to San Angelo in 1888. Short lines and branch lines extended like a fan to many cities on the plateau. The subsequent increase in population was phenomenal. No matter where the rails ran, settlers and sheepmen followed.

[13] Quoted in Ibid.
[14] Ibid.

New arrivals found little land available along favorable streams, so they pushed on to the high tableland along the divides. Here they built homes, grazed sheep, and for water dug wells. It was not infrequent in certain areas that digging for water resulted in dry holes. In such instances the pioneers built surface tanks to supply water for the greater portion of the year. These tanks, when properly constructed, held their capacity of run-off rainwater for a considerable time. Sometimes, however, water in tanks evaporated during protracted dry spells. Then ranchers erected windmills to pump water into the tanks. Throughout the Edwards Plateau there was sufficient wind velocity so that windmills could keep the reserve tanks well filled. But during summer months in drought years even windmills might not suffice.

Although the use of windmills on the Edwards Plateau began in the late 1870s, mass-produced factory-made windmills had been used elsewhere in Texas for perhaps two decades. As early as 1860 E. D. Nash of Columbia in Brazoria County advertised himself as the sole agent in the state for the manufacture and sale of Mitchell's Patent Self-Regulating Wind-Mills. Several other types of windmills could be found in Texas, but few of them were manufactured in the state. The U.S. Solid Wheel mill produced by the United States Wind Engine and Pump Company of Illinois was one of the first windmills used successfully in West Texas. The most common windmill on the plains of West Texas, however, was the Eclipse, made initially by a Wisconsin firm. The regular-pattern Eclipse windmills were made in sizes ranging from 8½ feet to 14 feet in diameter, but larger ones were also built. Another popular windmill on the Edwards Plateau was the Original Star, a solid wheel wooden mill, made by an Indiana Company.[15]

One of the first wool growers to erect a windmill was Christopher Columbus Doty. Born in Missouri in 1857, C. C. Doty came to West Texas in 1879, stopping at Uvalde, where his uncle W. T. Moore was in the sheep business. In the following years Doty herded sheep in South Texas, worked in San Antonio, and ran his own sheep operation on state land. In 1882 he established regular headquarters at the Ten-Mile Water Hole, located between Eldorado and Christoval. Here he built a picket house and corrals. He believed the water hole would provide a

[15]T. Lindsay Baker, "Turbine-Type Windmills of the Great Plains and Midwest," *Agricultural History* 54 (January, 1980): 38–51; T. Lindsay Baker, "Windmills of the Panhandle Plains," *Panhandle-Plains Historical Review* 53 (1980): 71–110.

permanent supply of the scarce resource. But in the summer the water began to dry up. Doty and his Mexican herders dug into the bed of the draw just below the water hole and struck plenty of water at a depth of only five feet. But the task of watering his four thousand sheep by hand proved too great. It seemed evident that another source was needed. Believing that by boring a deep well he could get an unlimited amount of water, Doty conferred with C. B. Foote at Ben Ficklin, then the seat of Tom Green County. Doty wanted to dig a well. Foote, who had the only well-drilling rig in that section of Texas, ridiculed the idea, suggesting that it was foolish to think of drilling on the dry upland range south of the Concho River. He further objected that he might get the drill hung in the underground rocks and have no end of trouble. He finally consented to dig the well, but indicated that he would charge twenty-five dollars to move the rig and one dollar per foot for drilling. Doty, calling the cost prohibitive, rejected the high bid and ordered a horse-powered drill from Fort Smith, Arkansas. It arrived by railroad at Abilene, and Doty hauled it overland to his headquarters. In July, eight days after returning and setting to work with the rig, he struck water. He found the water clear, pure, and cool. Over the well he erected an Original Star windmill and enjoyed more than sufficient water for his stock. As late as the mid-1940s the well was still in use.[16]

The use of windmills and tanks encouraged more settlers. As they arrived, people established more and more towns. As towns and cities appeared, warehousing expanded. Ballinger, Menard, Sonora, Eldorado, Eden, Del Rio, Mertzon, and Uvalde joined Kerrville, San Antonio, Encinal, Abilene, and San Angelo as important wool, and later mohair, warehousing centers. Agents handled wools consigned to these houses on a commission basis. Grading committees at each examined the wools, classifying them before they offered all or certain lots for sale, usually at sealed bids on designated dates. If a bid was satisfactory, the wool grower accepted it, otherwise he rejected it to offer the wool for sale at a later date. The warehousemen not only handled the wool and mohair, but also helped to finance ranching operations by making loans to the wool growers. Most notable lenders were T. C. Frost of San Antonio and Charles Schreiner at Kerrville, but others of-

[16] Newspaper clipping, Morgue Files, Texas Sheep and Goat Raisers Association, Records, Southwest Collection, Texas Tech; Roy Holt, "C. C. Doty, West Texas Pioneer," *Sheep and Goat Raiser* 22 (November, 1941): 21.

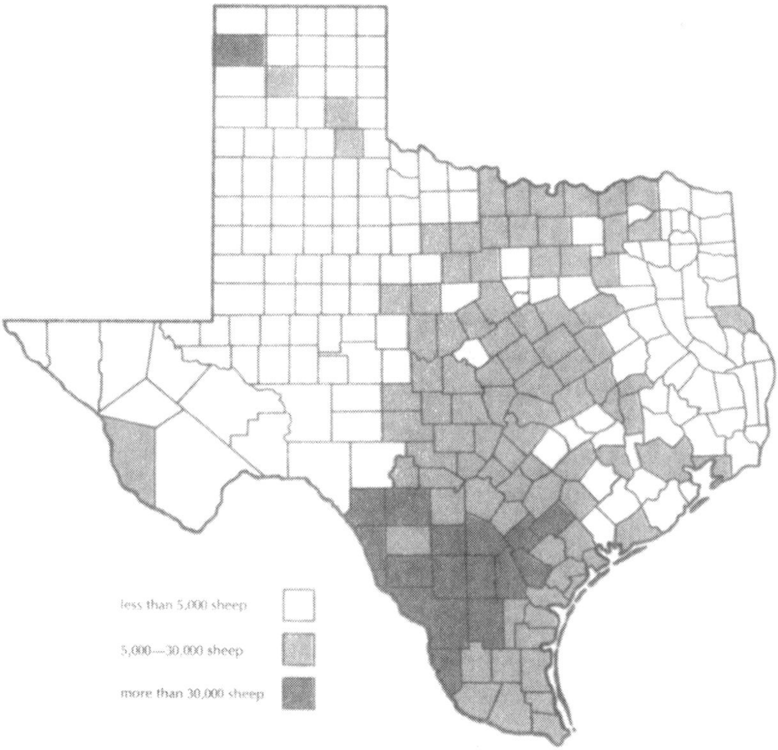

Distribution of sheep in Texas, by county, 1880

fered loans, too, usually accepting the wool or sheep as collateral. The Texas warehouse system was unique, no other state having such coverage or such widespread use of the method.

As the warehousing system developed, the numbers of sheep increased. There were, according to one account, about 376,000 sheep on the Edwards Plateau in 1880, representing about 16 percent of the state's total sheep population. By 1890 the number of animals had increased to 1,295,000, or 37 percent of the state's production, making the Edwards Plateau the largest area of sheep concentration.[17] Most of the sheep grazed in the counties below the Concho River south of San Angelo. The number of Spanish *chaurros*, however, declined. With

[17] Samuel Lee Evans, "Texas Agriculture, 1880–1930" (Ph.D. diss., University of Texas, Austin, 1960), pp. 268–71.

heavy emphasis put on long, fine wool, sheepmen upgraded their stock, emphasizing wool qualities that were not found in the *chaurro*. The Spanish animal lost favor, as Merino, Delaine, and Rambouillet ewes and crossbred ewes were obtained for their wool qualities. Later, when marketing of lambs proved profitable, English mutton sheep entered the plateau. The most popular mutton breed after 1900 was the Suffolk, characterized by a long, full body and by smoothness in its top and bottom lines. It was also a hardy, early-maturing sheep. The Suffolk, a highly distinctive animal with its jet black legs and face and white fleece, in the twentieth century was bred to Rambouillets, thus mating a strong mutton sheep with one of excellent wool qualities.

As a result of the widespread efforts to improve the quality of wool, the lambing season on the Edwards Plateau became a preeminent one. As a rule, the range ewes were bred in the fall—the gestation period about 150 days—to lamb between February and May, depending on the latitude and elevation of the range. The herder acted as midwife to each of the ewes in his flock. It was tough work. One participant called the lambing season a "month-long hell of worry and toil." The very future of the ranch depended on the number of ewes that gave birth to live lambs and the number of those lambs that could be brought through their critical first weeks. Wool growers often divided the pregnant ewes into "drop bands," watched them day and night, and when necessary attended them quickly, for during birth they were not only more defenseless than usual, but also more confused. Larger operations gathered for the task special crews, *hijadores* ("lambers"), composed of three or four men, including a cook and a nightman. The nightman walked among the bands with a lantern to find ewes ready to lamb. Finding one, he put her into a special wagon or a cage divided into compartments so small that it forced the mother to let the newborn lamb nurse.[18]

Young ewes caused the most problems. Indifferent mothers, they frequently failed to recognize their offspring and would wander off, oblivious to the lamb's tottering efforts behind them. One method of solving the problem was to tie the delinquent ewe to a nearby bush. Another method required the sheepman to hold the ewe between his knees and guide the lamb to its mother's udder, sometimes squirting a

[18] Ogden Tanner, *The Ranchers*, p. 107; Kupper, *Golden Hoof*, pp. 52–53.

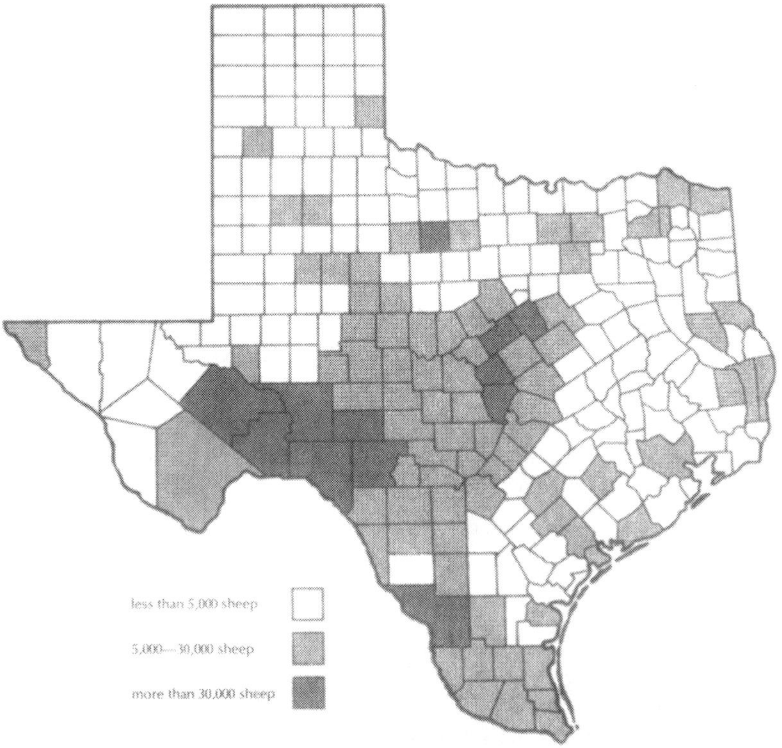

Distribution of sheep in Texas, by county, 1900

little milk in the lamb's face for encouragement. Occasionally lambs were born to "dry" ewes. In such a case herders tried to save the lamb by introducing it to a foster mother or feeding it a warm formula of milk, water, and molasses from a bottle. The care of orphaned lambs, known as bummers, was a major preoccupation for wool growers. The sheepman rubbed the bummer briskly with a rag until dry, wrapped it in a blanket or heavy cloth, and bedded it by a campfire. If a ewe with ample milk gave birth to a stillborn lamb, the herder attempted to get her to accept a foster offspring, whose real mother might be dead, dry, or overtaxed with twins or triplets. To accomplish the task required a stratagem called jacketing. The herder quickly took the pelt from the dead lamb, shook it right side out, and pulled it, sweater fashion, over the bummer's body. The smell of her offspring clinging to the jacket made it easier for the ewe to adopt the orphan as her own. After a few days, when the relationship was firm, herders removed the jacket.[19]

While it was toilsome, the lambing period was brief. The problem of wolves, coyotes, and wild dogs lasted all year through. The large gray or timber wolf, *Canis lupus*, inhabited western parts of the state. As early settlers destroyed their main prey (wild turkey, deer, antelope, small game), the wolves turned on the domesticated cattle, sheep, and goats. Their damage to sheep on the Edwards Plateau was awesome. When they attacked, they usually killed more than they could eat, leaving carcasses to rot or to be picked by scavengers. To protect the livestock, wool growers relentlessly exterminated the wolves or forced them farther west. When bounties were offered for wolf pelts, farmers and ranchers hunted them to near extinction.

The coyote, *Canis latrans*, was not so easily run off. This wily animal has adjusted unexpectedly well to modern civilization and man's encroachment on his domain. A little animal, standing about sixteen to twenty-one inches at the shoulder, it has a bushy tail and erect ears. The coyote eats rats, gophers, rabbits, and dead animals. Since it also kills and eats sheep and calves, farmers and ranchers on the Edwards Plateau have hunted it, with mixed success, for a century. Conservationists, however, maintain that coyotes do more good by destroying rodents and rabbits—which feed on the same grass as sheep—than they do harm as killers of livestock. On the plateau they have been hunted by guns, traps, poison, and planes. Recent efforts by wildlife experts on behalf of animal conservation have resulted in a government ban on poisoning coyotes, once the most effective means of controlling predators.

Except in the Hill Country, where collies were admired, dogs were not popular among Edwards Plateau sheepmen. While *pastores* used them with great success in the Rio Grande Plain, in the Panhandle, and in New Mexico, only a few wool growers used them on the plateau. Most Anglos had neither the patience nor the skill to train a good sheep dog. In Southwest Texas they were a nuisance, causing more harm than good. Too often the dog would be out chasing a rabbit or he would run the sheep too hard, bunching them up when they needed space to graze the sparse grass and making them nervous. Wild dogs, although not numerous in the earliest days, were even more

[19]Tanner, *Ranchers*, pp. 105–107.

troublesome. Running in packs, they struck savagely. Like wolves, the dogs destroyed far more sheep than their bloated bellies could hold. Less fearful of men than their shy cousins, the wild dogs tended to remain close to flocks when not hunting—waiting, it seemed, for another opportunity to strike. Authorities estimated that dogs killed 52,500 sheep in 1890.[20] Bobcats, mountain lions (cougars or pumas, called panthers by early settlers), eagles, and buzzards also attacked sheep.

Range fires were likewise destructive. In the early 1880s fire was an annual summer event. The grass, dried to hay by the hot sun, burned quickly, blackening the earth, charring the bear grass, and setting aflame mesquite trees. As it raced across the plains, the fire jumped dry creeks and scorched earth firebreaks. Sparks leaped into the prairie wind, turned with the breeze, and like a torch set another blaze. As often as not the fire ran uncontrolled, destroying a million acres of grass down to the roots and wiping out the herder's plans to graze his flock through the winter. There were many such fires, with few roads, trails, streams, or natural barriers to stop them when they once got under way.

One of the worst fires occurred in February, 1882. It started southwest of present Eldorado and swept over the whole region, reaching to the South Concho and to the head springs of the San Saba. The flames rushed across the plains in a wall of fire. As the fire neared his headquarters at Ten-Mile Water Hole, C. C. Doty moved to save his picket house and windmill. He and two hired hands were joined by M. L. Mertz, who also ran sheep in the vicinity, and a companion. After saving the house and windmill and driving Doty's sheep into a dry lake bed, these five men began to fight the fire at Eight-Mile Draw, northwest of present Eldorado. They fought all night. Armed with cedar brush, sacks, or slickers, they beat out the flames as they followed along the edges, always working toward the front of the fire. By daylight they had made some headway. During the following day the fire continued eastward, the wind taking it beyond control. The men prepared food and slept some during the day. On the second night, they fought it again, setting back fires to save some six sections of grass. For three more nights they continued the choking, stifling work. On the morning after the fifth night they met some men from Dove

[20] Evans, "Texas Agriculture, 1880–1930," p. 271.

Creek who were coming to their aid. It was too late. Practically all the range from the South Concho to the Pecos had been destroyed. Doty had worn out four new slickers during the fire and lost a pile of 15,000 cedar posts, which he and his help had prepared for fencing.[21] There was not one thing left in the wake of the fire except an occasional large mesquite tree. In July, 1884, another fire raced across the Edwards Plateau. This time grass burned from the Devils River to the Concho.

The woes of fire, drought, varmints, and low wool prices took their toll on the Edwards Plateau sheep industry. The number of sheep in the vast region declined after the 1880s to only about 642,000 head in 1900. New tariff laws that increased foreign wool supplies in the United States and the nationwide economic depression in the 1890s were in part responsible. Hundreds of men left the country, abandoning their leased land or selling out. The population of the plateau stabilized, or even declined slightly, as men looked for opportunities elsewhere. Some communities failed—or like Ben Ficklin were swept away by flood. Only the hardy souls stuck it out in the sheep business during the early 1890s. One of those who did was Sam M. Oglesby, a rare character who ran sheep on the Edwards Plateau continuously for nearly fifty years—a record of some sort.

Oglesby, born in Lynchburg, Virginia, came to West Texas in 1883 with ten dollars in his pocket. A young man with a slender but well-set body, flashing dark eyes, and a polish little known in the rugged sheep country, he attracted attention. A brickmaker by trade, he turned to that occupation in San Angelo. Then he worked on the Leasel Harris ranch near Robert Lee in Coke County. After his marriage to Donna Blanks in 1887, he farmed on some river bottomland for a year before going to work on the sheep ranch of his father-in-law at Sherwood. Wanting a business of his own, Oglesby bought a livery stable at Sherwood in Irion County, but three years later he sold it to Ben Cornelison and returned to wool growing with Bob Dameron. That unhappy partnership lasted just over a year, ruined by the 1890s depression. Oglesby again bought into the livery business. Finally, in 1894, he bought a ranch in Schleicher County, just across the Irion County line. He introduced sheep immediately. As the years passed, he increased his flocks and added to the original holdings by lease and purchase, obtaining property near Sherwood and in 1910 buying a small ranch ad-

[21] Holt, "C. C. Doty, West Texas Pioneer," p. 28.

jacent to Mertzon. This one he made his headquarters and home.[22]

As Oglesby and others struggled through a period of low wool prices, the Angora goat and mohair industry claimed attention. The work of William L. Black, Charles Schreiner, J. D. Pepper, and W. D. Parish was in part responsible. But wool prices that were temporarily down, the interest mohair was attracting among textile manufacturers, and the fact that the Edwards Plateau afforded a variety of the best Angora browse to be found in the United States were additional factors accounting for livestockmen's growing concern with Angora goats. Besides, sheepmen found it easy and profitable to raise Angoras. Most of them already ran a few common Mexican (or Spanish) goats with their sheep, and, thus, the switch to Angoras could be made at minimal cost. There were perhaps fewer than a half-million goats of all kinds in Texas in 1880, most of them the Mexican variety. Through the next decade their numbers barely increased, but in 1900 Texas contained some 627,000 animals, over 75 percent of them concentrated in the upper Rio Grande Plain and the southwestern part of the Edwards Plateau. Of these, some 100,000 were Angoras or Angora crossbreds.[23]

The Angora industry developed along lines similar to the sheep industry. Indeed, from the first, mohair growers were sheepmen, too. Only rarely in the early days did someone on the Edwards Plateau confine his livestock interest to goats. Mohair growers acquired purebred bucks to breed to common does in much the way sheepmen had upgraded *chaurros*. Shearing goats was accomplished in much the same manner as shearing sheep, although it was the general practice to shear goats twice a year—in the spring and fall. Since cold, wet weather immediately following shearing could cause heavy losses, some ranchmen left a "cape," a strip of mohair three to four inches wide, down the back of each goat. The long hair on the cape, hanging down on each side of the body offered some protection from bad weather. Others sheared with raised combs, leaving about one-quarter

[22] Newspaper clippings, Morgue Files, Texas Sheep and Goat Raisers Association, Records, Southwest Collection, Texas Tech; San Angelo *Weekly Standard*, June 26, 1937.

[23] George T. Willingmyre et al., *The Angora Goat and Mohair Industry*, Misc. Circular No. 50, United States Department of Agriculture, 1929, p. 11; T. R. Hamilton, "Trends in Production, Use and Prices of Texas Mohair," *Sheep and Goat Raiser* 25 (July, 1945): 14–17, 42–45; William Leslie Black, *A New Industry—Or Raising the Angora Goat*, p. 104; Agricultural Marketing Service, "Livestock Numbers in Texas," *Sheep and Goat Raiser* 22 (February, 1942): 12–16; *Texas Almanac*, 1980–81, p. 577.

of an inch of hair on the goat. After it was shorn and packed, mohair was transported to warehouses to await inspection by buyers representing eastern mills. Processing mohair into a finished fabric was very similar to processing wool. Mohair was graded, scoured, combed, spun, woven or knitted, and washed, just as wool was.

While several prominent sheepmen entered the Angora industry in the 1880s, an even bigger spurt came in the next decade. Judge Bob Davis of Uvalde County became involved in 1893, when cattlemen gave him four calves. Davis traded the calves for forty Angora does. Over the next few years he set about building up one of the finest Angora bands in that part of the state. His sons, Arthur and Bob Davis, Jr., carried on his work in the Uvalde area. The Davises were instrumental in having the headquarters for the American Angora Goat Breeder's Association moved to Sabinal, Texas, from Kansas City in 1924. In 1926 the association headquarters was located at Rocksprings in Edwards County, where it remains today. Claude Pepper, of Edwards County and son of J. D. Pepper, who had established a successful band in the 1880s, enjoyed one of the oldest continuing Angora lines in the country. James Prentice of Kimble County, who would win many prizes for his purebred Angoras at stock shows in the twentieth century, was another of the first men to run goats in West Texas. He was a charter member of the American Angora Goat Breeders Association.[24]

Thus, by the mid-1890s the Edwards Plateau had become the state's leading region for the production of sheep and wool. Its farmers and ranchers had fanned out over most of the vast tableland, driving sheep before them as they introduced new breeds, struggled with predators and fires, erected windmills, and made the isolated and empty region a viable and productive part of Texas. By that time the Edwards Plateau had also become the center of an emerging Angora goat and mohair business. Although the sheep industry was struggling through a brief period of retrenchment at the close of the century, Angoras were only beginning to attract ranchers as a possible supplement to sheep and cattle raising. The industry, like the sheep and wool trade half a century earlier, needed only a promoter. It found one in William Leslie Black.

[24] Newspaper clippings, Morgue Files, Texas Sheep and Goat Raisers Association, Records, Southwest Collection, Texas Tech.

8
William Leslie Black

MANY ranchers, including W. W. Haupt, Joseph P. Devine, W. D. Parish, and others raised Angoras in Texas before him, but nobody did more to promote and encourage the Angora and mohair industry in the state than William Leslie Black. An intelligent and enterprising individual, Black experimented with Angoras, their mohair and meat, and encouraged widespread adoption of the animal for livestock operations on the Edwards Plateau. He popularized the breed, sold his animals outside the state, established canning and tanning factories, and promoted the goat enthusiastically in letters, newspaper articles, pamphlets, and books. From shortly after the time this prominent author and goat raiser entered the business, Texas became the number one producer among states of Angoras and mohair.

Born in New Orleans in 1843, William L. Black was of Scottish and English ancestry. His father, a successful middle-class businessman who had migrated to America from Scotland, owned a summer home on Lake Pontchartrain where young Black and his brothers, Charles and Alexander, enjoyed many exciting summer outings, fishing, swimming, and rowing on the lake. Bright and resourceful, William left school at an early age to work at his father's merchandising firm. Here, Black gained experience in overseas cotton trade, a business he one day made his own.[1]

In 1862, during the Civil War, at age nineteen William and Charles enrolled in the Confederate Army. Their volunteer regiment, called the Crescent, signed up for the standard ninety days. Shortly afterward, with little training or preparation, the recruits accepted as-

[1]William L. Black, "Autobiography," typescript, William L. Black Papers, Southwest Collection, Texas Tech University, Lubbock.

signment in Mississippi, where they served under General P. T. Beauregard. Two days after arriving in Beauregard's camp, they fought in the sanguinary battle of Shiloh. Both William and Charles sustained serious injuries.

After the bloody battle, to recover from their wounds, the Blacks returned to New Orleans. Their father moved the family to its summer home during the convalescent period and arranged an honorable discharge for his boys. Because his mother was distressed about his serving in the army, William promised not to reenlist. A true southern son, however, he refused to be a shirker, as men of fighting age who did not enlist were called. Deciding to participate in other ways, he went to Augusta, Georgia. There his brother Alexander shipped cotton to Great Britain via Nassau in the Bahamas, running the Union blockade of the southern coast. Not long afterward when a shortage of men arose, William accompanied a shipment to Nassau.

In the Bahamas Black met Lieutenant Thomas E. Hogg of the Confederate Navy. Hogg had been commissioned to fit out an expedition to capture either the steamer *Guatemala* or its sister ship the *San Salvador*, both of which made semimonthly trips between Panama and Guatemala on the Pacific coast. Once in command Hogg would use the pirated vessel to capture two large passenger steamers, which were then the only means of Union communication between Central America and San Francisco, destroy the Pacific whaling fleet, and terrorize United States commerce, sending the spoils to the Confederacy. He offered Black the position of paymaster with his crew. Young Black, believing he was not violating his promise to his mother about enlisting again in the army, in 1864 accepted an appointment in the Confederate Navy.

The Hogg expedition failed in its mission. Black and six of his companions were captured by Union troops while on board the *San Salvador*, taken to California, and incarcerated at Fort Alcatraz in San Francisco Bay. The efforts of D. J. Tallent, a San Francisco banker, secured Black's release in December, 1865, but Black remained in California until May, 1866, when his companions were liberated.[2]

In the years following his prison release William Black kept busy. He entered the wholesale fruit and nut business in New Orleans.

[2]"Scrapbook," Black Papers.

When that failed, he considered operating a fishing fleet in the Gulf of Mexico but settled instead on entering the cotton trade, first in New Orleans and then in New York. He acquired considerable wealth, established with others the New York Cotton Exchange, traveled widely, married, and in 1875 moved to Saint Louis.[3]

Meanwhile, Black turned his attention to wool. Interested in the commodity as a more profitable proposition than cotton, he toured West Texas, especially the Edwards Plateau country scouting the region for an ideal place to raise sheep. "In 1876," he wrote afterward," I made a location of 30,000 acres of land at the headsprings of the San Saba River in what was supposed to be Crockett County, but was afterwards made part of Schleicher County." He subsequently purchased the land, which is in present Menard County, from Robert Robinson for ten cents an acre, but even at that price his new friends in Saint Louis thought him mad. Black had another opinion: "I thought," he remembered, "with the splendid spring I got with my purchase, and a 300-acre grove of fine pecan trees, it was reasonably cheap."[4]

The West Texas land was located near Fort McKavett, 180 miles northwest of San Antonio. The fort was part of a chain of garrisons, including Forts Duncan, Clark, Concho, Griffin, and others stretching from the Rio Grande to the Red River, designed to secure the westward-advancing frontier against Indian attacks. Established in 1852, Fort McKavett was garrisoned by the United States Army until 1859, when the troops withdrew. During the war fewer than one dozen families in the neighborhood kept guard over their flocks of sheep and herds of cattle. After the war, it was reoccupied by federal troops and maintained until 1883, when the army again abandoned it because it was no longer needed.

Not yet ready to move to West Texas, in 1876 Black hired Douglas Shannon, an English remittance man who knew no more than his employer about the sturdy demands of a stock ranch, to take charge of the Texas property. Neighbors and some of the employees, like Charlie Adams, who later helped to found Sonora, thought Shannon an odd sort. Among other things, Shannon nailed up all over the forty-six-

[3]Ibid. See also Edith Black Winslow, *In Those Days, Memories of the Edwards Plateau*, pp. 4–10.

[4]William L. Black, "Ranching on 10¢ Land in Texas," *Cattleman* 14 (July, 1927): 31. See also "The Saga of Colonel Black," *Sheep and Goat Raiser* 30 (July, 1950): 3–5.

section spread copies of "Ranch Rules and Regulations." The broadside admonished his men that they must be early to work at all times, avoid alcohol, and not leave their task without permission. Shannon also bathed daily in the river, even in winter when ice was thick.[5] Shannon, if uninformed about ranch life and the ways of Texas cowboys and sheepherders, soon learned to handle ranching operations. He bought a flock of one thousand *chaurros* for the standard one dollar per head, a train carload of purebred Merino rams, and a herd of seven hundred cattle, paying for the latter five dollars each. He erected barns and windmills, built cabins for the ranch hands, and made other improvements. The Black-Shannon arrangement called for the remittance man to receive one-third interest in the profits of the ranch for five years. The bargain proved a good one for both men.

At least once a year Black went out from Saint Louis to inspect his ranch. After each trip his desire to establish a permanent home in Texas increased. His wife was not enthusiastic. Edith, whom Black had met in Nassau during the Civil War, was the daughter of George Clark Bogert, a well-to-do businessman from New Orleans, who had fled to Nassau after Union troops occupied the city. Born and reared in the city, Edith was used to urban comforts and surroundings and not at all sympathetic to country life or pioneer living. The couple had seven children, and Mrs. Black wanted them to remain in city schools. But in 1884 she relented, and the family moved to Texas.[6]

The trip from Saint Louis was a pleasant one. The family traveled by rail to Abilene, Texas, about 125 miles from Fort McKavett. From there they went by wagon. The Blacks, with three women employed as household servants and the family dogs, rode in two army ambulances, wagons that contained two seats and canvas curtains that could be rolled down in bad weather. Built high off the ground to ease fording of rivers and streams, the ambulances provided fairly comfortable travel. Behind came seven covered wagons, containing part of the household goods and furniture. Although it took several days to make the trip, Black found suitable lodging for the party each night. At Fort

[5] Roy Holt, "Pioneering Sheepmen," *Sheep and Goat Raiser* 26 (December, 1945): 20–21; Winslow, *In Those Days*, pp. 14–15.

[6] Winslow, *In Those Days*, pp. 3, 11–13, 17–20; Kate Adele Hill, *Home Builders of West Texas*, pp. 58–70.

McKavett, about a mile and a half from the headquarters ranch, the family lived in the hospital until the ranch house was completed.[7]

In Texas Black prospered. Upon his arrival, he sold his cattle at a good profit and entered full-scale into the sheep business. When Shannon moved on, Thomas (Tom) Palmer and A. B. Priour became foremen of the ranch. Since most of the employees were of Mexican descent, Black continued, with modifications, the traditional Spanish system of sheep management, with its *pastores*, *vaqueros*, and *caporales*. In spite of the prosperity, sheep raising was not without its problems. For Black the most pressing one was wool prices, which had dropped from twenty cents a pound in 1884 to seven cents in 1885, when he sold his first clip in Abilene, where Theodore Heick had just opened a wool warehouse. Because of the severe drop in wool prices, Colonel Black, as he was called in Texas, restocked his ranch with cattle.

Unwilling yet to abandon sheep, Black rented from the state at the rate of three cents an acre per year an additional 30,000 acres adjoining the nearly fifty sections he had purchased in 1876. Around all of this he erected the first barbed-wire fence built in the county. Including fence posts, wire, and labor, the cost was enormous, but, Black reasoned, necessary. A small herd of wild horses was shut in at the time of the fencing. Several years later Black's eldest son, George, and foreman Palmer rounded up a load of mavericks, hauled the animals by train to Missouri, and there sold them at public auction. Black grazed his sheep on the rented property, called the Point Breeze Ranch, putting Priour in charge.

Black became active in a local wool growers association. He participated in its irregular meetings, served on its committees, and took a leadership role in the organization. In 1887 he represented Texas at the National Wool Growers Association meeting in Saint Louis. The major interest that year concerned the efforts of the assemblage to pressure the United States Congress to raise the duties on imported wool. The success was mixed. While it did not raise the tariff duties, Congress did not include wool on the list of free materials, as Black expected.

[7]Winslow, *In Those Days*, pp. 17–20.

Two years later, Black, expecting wool prices to increase, purchased 8,000 head of sheep. These he bought at Knickerbocker, south of San Angelo, for seventy-five cents per head. To the flock he bred improved Merino rams. When he bought the sheep, Black hired C. C. Doty, the West Texas pioneer credited with having drilled the first well on the upper Edwards Plateau. Trouble with wolves, cattlemen, drought with its expected fires, and low wool prices had forced Doty to sell out in 1884. Two years later he took charge of 7,000 sheep on shares and in 1889 joined Colonel Black as assistant to Priour at Point Breeze Ranch. Doty stayed until 1893, when Black again sold his entire flock of 20,000 sheep.

Colonel Black, in the meantime, entered the Angora goat business. As early as 1876, when he bought the West Texas property, he learned that Mexican herders perferred to eat goat. Since goats were cheaper than sheep, he purchased a small flock to supply his herders with fresh meat. By 1884 he had discovered that Angoras, gaining popularity farther east, were a superior breed and possessed greater commercial value than the common Mexican species he owned. Knowing that he could upgrade his common stock by introducing full-blooded Angora sires, he bought from Colonel Richard Peters of Atlanta, Georgia, eight males and four females, paying $750 for the goats, including delivery to his ranch. Although he considered it an enormous price for goats, the investment proved a solid one. In a few years Black increased his flock enough to justify shearing the animals to sell the mohair. Within eight years of purchasing the purebreeds he counted nearly eight thousand head of well-graded Angoras.[8]

Black found the animal perfectly adapted to his San Saba River country. The altitude and climate were ideal, and he had an abundance of fresh, clear water. The hilly range along the breaks of the San Saba provided year-round browse, and the Angoras helped to clear the brush from his cattle pastures. His nannies produced young at a rate of nearly 125 percent per year. Neither disease nor scarcity of grass depleted the flock, for the animals lived on the leaves and shoots of scrub oak, mesquite, and other shrubs of the region. Nor did the crossbreeds require close watching. They took care of themselves, regularly came home with a herder to the pens at night, but did not require close

[8] William Leslie Black, *A New Industry—Or Raising the Angora Goat*, pp. 12–13.

stables. Black's crossbred flock doubled in size each year, and after eight generations of breeding he had difficulty distinguishing between his purebred and crossbred Angoras. Moreover, he experienced no problem in selling the mohair. Wool warehousemen welcomed it as they did wool. Textile manufacturers liked its silky luster, its great affinity for dyestuffs, and its strength and durability. They used it extensively in the manufacture of drapery, portiere cloth, linings, cloaking, knitted garments, rugs, and numerous other fabrics. Manufacturers of railroad passenger cars preferred mohair to wool for upholstery.

In 1893, when a devastating panic and depression smothered the United States, Black sold his cattle. He also mortgaged the greater part of his land. Worrying over a new and unfavorable tariff law, Black also sold his 20,000 sheep. Thus, he escaped another seven-cent-per-pound price era that resulted partly from the depression and partly from, as Black had feared, a new tariff that allowed foreign wool to enter the United States without duty. He still browsed 8,000 Angoras. These, he figured, must be thinned out, or in a couple of years they would overrun his ranch.

Accordingly, Black contacted a personal acquaintance with Armour and Company, a Chicago meat-packing firm, offering to sell one thousand fat wethers at Armour's price, if it would return the hides. Declining, the friend informed Black that goat meat was not popular among Armour's customers and did not sell. Nor would the company venture to buy the animals in large numbers until people had overcome their prejudice against it. His friend advised slaughtering the goats for their hides and tallow and packing the meat in cans labeled "roast mutton." Deciding there was money to be made in such a dubious proposition, Black concluded that he would butcher the animals himself.[9]

Thus was born the Range Canning Company, one of West Texas' first meat-packing firms. It was also a rendering plant and a tannery. Black, after returning from New York, where he had gone to obtain investors and money from old friends, dammed up the San Saba River and built a waterwheel to provide power. He constructed a large frame combination slaughter house and cannery with a rendering plant

[9]Ibid.; *San Angelo Standard-Times*, August 29, 1954. See also *Texas Livestock Journal* (Fort Worth), October 28, 1892.

nearby, bought machinery, purchased thousands of two-pound cans, and ordered beautiful four-color labels. He hired W. G. Tobin from a Chicago meat-packing company to oversee the work. He built the tannery on a low bluff overlooking the dam, brought in an experienced tanner, and employed about twenty-five people, mostly Mexican-Americans, to carry on the canning, tanning, and rendering operations. To house his employees, Black erected a row of small homes and established a two-story store for workmen and their families. The ranch now resembled a small factory town.[10]

Black slaughtered three thousand old nannies and wethers in 1893 and the following year four thousand other animals. In 1895 in addition to Angoras he slaughtered hogs. The canning operations went smoothly, but Black found difficulty selling his so-called roast mutton. When new labels, calling his meat boiled mutton, did not help, he changed the labels again. This time he printed them as "W. G. Tobin's Chili Con Carne." That spurious scheme worked. He contracted to sell his product to a Chicago packer who, perhaps typically for some meat packers of that day, disposed of the mislabeled product in Europe. Black apparently thought well of the deception. He made some money on the venture. As for the tallow, not only did he find a market for it, but the broker with whom he placed it asked for more, reporting that it was of a superior character and would command a premium for making fine candles. In January, 1895, Black shipped 20,000 pounds of tallow to Chicago. Altogether, he realized from the sale of canned meat, lard, and tallow a profit of almost three dollars per animal.[11]

The tanning operations were also profitable. Black experienced little trouble in finding a ready sale for his dressed skins, selling as many as one thousand to a single wholesale house, Marhall Field's, in Chicago. Most of the skins were left in their natural color, but others were dyed pink, blue, yellow, or green. The success of the tanning business encouraged nearby ranchers to bring in hides of bobcats, coyotes, jackrabbits, and other animals to be made into warm carriage robes. Black's workmen lined the robes with a soft felt as broad as the

[10] Newspaper clipping, Morgue Files, Texas Sheep and Goat Raiser Association, Records, Southwest Collection, Texas Tech; Winslow, *In Those Days*, pp. 40–41; James Cox, *Historical and Biographical Record of the Cattle Industry and Cattlemen of Texas and Adjacent Territory*, pp. 356–57.

[11] Black, "Ranching on 10¢ Land," pp. 31–32; Black, *A New Industry*, pp. 13–15; *San Angelo Weekly Standard*, January 12, 1895.

robe itself. On some of the robes they left the tails of the animals, thereby enhancing, for some unexplained reason, the value of the article. Some area ranchers brought the skins of lambs that had died to be fashioned into exquisite baby slumber robes with pink or blue quilted satin used in the lining. Some workers manufactured gloves. So busy was the enterprise that Black used the upper floor of the store as a sewing room in which to turn the pelts into robes. For this work he employed women in the neighborhood and asked his wife to supervise the operation.[12]

By 1896 his goats had again increased to a large number. But Black determined that rather than resume his slaughtering operation, he would sell his surplus animals to area stockmen interested in grazing Angoras. That was not hard to do. The Edwards Plateau was a natural habitat for goats, and ranchers there were realizing that Angoras were easy to raise and that mohair production was an attractive proposition. Accordingly, Black promoted the sale of his goats. In response, ranchers invested heavily in Angoras. Black sold thousands of goats in some two dozen area counties. For a brief time, as he labored to cut his flocks, his ranch became a busy auction mart and trading center. As a result of all this activity, Black's goats were the foundation of many West Texas flocks.

Several people from outside the state visited Black. One man, Dr. J. R. Standley of Platteville, Iowa, wanted to ship goats northward to sell at public auction. Indicating that he had cleared about five hundred acres of his farm by running goats, Standley offered to buy all the goats Black would ship. Caught up in Standley's enthusiasm, but not willing to ship all his goats, Black in 1897 entered into an agreement to deliver 1,000 goats for sale on Standley's farm. At the sale, occurring in July, farmers bought them all. That fall Black sent a second shipment of 1,000 head. The following year he delayed shipping until buyers of his goats had sufficient time to test the value of the animals. But when demand increased, he and Standley resumed the enterprise. In fact, sales were so brisk that Black found it necessary to buy from area raisers additional Angoras for shipment to Iowa. By 1900 he had sold more than 15,000 Angoras in the Hawkeye State and was selling them elsewhere as well.[13]

[12] Winslow, *In Those Days*, pp. 40–41.
[13] Black, *A New Industry*, p. 15.

The Angora goat industry spread (see tables 4 and 5). With few exceptions every state in the country had Angoras at one time or another, but in 1900 Texas led the nation, producing over one-third of all Angoras in the United States. Her farmers and ranchers grazed some 627,000 goats. Of these, some 100,000 were Angoras, which when clipped produced an estimated 275,000 pounds or more of mohair classified as "mostly short, six months' staple, but very clean and light fleeces, fluffy, and cottony."[14] Twenty years later Texas counted nearly 2,000,000 goats and accounted for over 75 percent of the country's total mohair clip. In Texas the Edwards Plateau became the center of the industry, completely dominating the state's output.

By the end of the nineteenth century Black had spent countless quiet hours studying the nimble little animal whose curly locks hung down all over its body. He had read all he could about Angoras, stopping at bookstores while on business trips as far away as New York. He had observed the animal, its habits, and forage. He had made careful notes about its routines. Particularly he had become excited over its apparent resistance to diseases that wasted his sheep and cattle. In the 1890s he published a little pamphlet on the Angora, setting forth some of the advantages of the goat industry and calling upon farmers and ranchers to seize the opportunity offered them in what he called "this new, and valuable, industry."[15] In response he received scores of inquiries, asking all manner of questions: whether Angoras could survive in cold climate, the kind of fence needed to hold them, and others. He answered most of the letters at some length, explaining various details of Angora and mohair production. He tried to convince each writer of the economic prospects in goat raising.

So convinced was Colonel Black of the good future for Angora goats that he wrote a book on the subject: *A New Industry—or Raising the Angora Goat*. In the lengthy treatise Black dealt with the industry's history as well as its commercial and practical features. He talked

[14] Ibid., p. 104; Texas Crop and Livestock Reporting Service, *Texas Historic Livestock Statistics, 1867–1976*, pp. 35–39; George T. Willingmyre et al., *The Angora Goat and Mohair Industry*, Misc. Circular No. 50, United States Department of Agriculture, 1929, pp. 11–19; U.S. Deparmtent of Agriculture, Division of Crop and Livestock Estimates, "Report," typescript, May 5, 1923, Correspondence File, American Angora Goat Breeders Association, Records, Southwest Collection, Texas Tech.

[15] Black, *A New Industry*, pp. 1–10, 15.

TABLE 4
All Goats, Mohair Goats Clipped, and Pounds of Mohair in Texas, 1900–30

Year	All Goats	Mohair Goats Clipped (No. spring-clipped plus fall-clipped kids)	Pounds of Mohair
1900	627,000	135,000	275,000
1910	1,535,000	1,320,000	4,092,000
1920	1,998,000	1,834,000	6,786,000
1930	3,117,000	3,518,000	14,800,000

SOURCE: Texas Crop and Livestock Reporting Service, *Texas Historic Livestock Statistics, 1867–1976*, pp. 36–38; U.S. Dept. of Agriculture, "Agriculture Statistics," typewritten release, 1943, in Correspondence Files, American Angora Goat Breeders Association, Records, Southwest Collection, Texas Tech, Lubbock; T. R. Hamilton, "Trends in Production, Use and Prices of Texas Mohair," *Sheep and Goat Raiser* 25 (July, 1945): 14; Natural Fibers Economic Research, *Texas Sheep and Goat Wool and Mohair Industry*, pp. 9–10.

TABLE 5
Annual Gross Income from Mohair Received by Producers in Texas, 1909–20

Year	Income	Year	Income
1909	$1,061,000	1915	$1,525,000
1910	1,159,000	1916	2,457,000
1911	1,375,000	1917	2,503,000
1912	1,448,000	1918	3,447,000
1913	1,428,000	1919	3,483,000
1914	1,347,000	1920	1,825,000

SOURCE: T. R. Hamilton, "Trends in Production, Use and Prices of Texas Mohair," *Sheep and Goat Raiser* 25 (July, 1945): 43.

about manufacturing uses of mohair. He provided biographical information on early Angora breeders, calling Richard Peters of Georgia the "Father of the Angora Goat Industry." He included notes from breeders, giving their experiences in handling the animal, and letters he had received from Angora raisers concerning the utility of the breed. He stressed the idea of using goats with sheep and cattle to form a Texas ranching triumvirate, indicating that in most ranges of the Southwest sheep, goats, and cattle could pasture together. Cattle, he noted, ate tall grasses, sheep nibbled on short grass, and goats preferred the leaves of trees and scrubs. Finally, he offered the book as a manual on the care and management of Angoras, with information on upgrading the common Mexican goats. The book was a popular success. Until it went out of print many years later, the study was regarded as the standard work on Angora goats.

Black never remained long in one enterprise. He was basically a promoter. After starting something, he abandoned it to others and pursued different interests. For example, he never resumed the canning operations, although he claimed they had been profitable. He ran cattle, sheep, and goats on his San Saba River ranch. But while he endeavored to promote the Angora industry in the state, goats to him were always of secondary economic importance. Early in the twentieth century he turned the ranching operation over to his sons and studied other matters. He invented a cotton picker. He patented a "special guide" for windmills. He produced what he called his Little Wonder Nut Sheller, which he had developed especially to use on the pecans growing on his property. He participated in the Texas Sheep and Goat Raisers Association. He continued his writing. And he sought with success the creation of a futures market for wool and mohair. He was eighty-eight years old when, four weeks after a paralytic stroke, he died in 1931.[16]

Clearly the saga of William Leslie Black is remarkable. At one time or another, he was a daring adventurer, a Confederate pirate, a Union war prisoner, a wealthy cotton broker, a charter member of the New York Cotton Exchange, a cattle and sheep raiser, a successful inventor, a productive writer, and a dubious meat packer. More than any other person, he was responsible for the growth of the Texas Angora goat and mohair industry. While Black labored to build sheep, cattle, and goat operations along the San Saba, others engaged in the sheep-trailing industry.

[16] Newspaper clipping, Winslow Family Papers, Southwest Collection, Texas Tech.

9

The Sheep-Trailing Industry

THE trailing of sheep and goats is as old as the industry itself. Methods developed in the ancient Near East were improved upon in the Middle Ages in Europe, when Spanish dons developed the *transhumante* system. Sheep- and goat-trailing spread with the animals to other parts of Europe before coming to America. In regions that later became part of the United States sheep and goats arrived with the first Spaniards. Later Anglo-Americans migrating westward took sheep with them to unsettled frontier portions of the country.

The first sheep trailed into Texas were the *chaurros*, long-legged, hardy, and able to walk great distances. For two hundred years they dominated southwestern flocks. After 1800, when improved strains entered the region, farmers and ranchers upgraded the *chaurro* stock as they expanded rapidly across the state. *Chaurros* moved northward out of the Rio Grande Plain to the Hill Country, the Edwards Plateau, the Trans-Pecos, or beyond to states in the Rocky Mountains. Merinos, Southdowns, and others from fertile Midwestern farms, Bakewells and Cotswolds from the humid Old South, Delaines from New England's distant hills, Rambouillets from sprawling California ranches, and other sheep from Europe entered Texas to mix with the *chaurros*. Drovers trailed sheep everywhere. It was a minimally organized procession, however, which advanced and changed with the sheep industry itself.

Trailing sheep was not unlike cattle-trailing. Sheep-trailing lasted over a longer period but took fewer men. Sheep needed fresh water, moved slower than cattle, and were more vulnerable to danger from predators, lightning, fires, and Indians. Water crossings, too, posed greater problems for sheep trailers than cattle trailers. At large rivers and streams drovers commonly ferried their animals across or found a bridge, at which they often had to pay toll charges. The sheep could subsist on scarce grass, seldom developed hoof problems, and could be trailed in large numbers. Sometimes sheep moved overland in bands of 10,000 to 25,000 head, but usually the number was smaller. Sheep tended to spread out farther than cattle on open range. A flock of several thousand might stretch over a five-mile front, gathering together only at the bed-ground site. In semiarid portions of Texas it was not uncommon for drovers to bed their sheep by day and move them during the relative cool of the early morning and late evening. Wherever they might be heading, sheep with rare exception fattened and throve as they moved, for grazing from range to range was a natural inclination for them—the pattern of their life.

Sheep drovers were unique characters. Most of the successful ones among them had learned the business from the bottom up, living near the animals at all seasons. They have been described as quieter, steadier, and more careful than cowboys. Sheep trailers displayed such qualities as stability, innovation, open-mindedness, and a tolerance of uncertain situations. They have often been pictured as peaceful, docile, and lawabiding, but more often their life was adventurous and full of obstacles. They were conscientious and hard-working and above all loyal to their flock of exasperating animals.

Drovers often had difficulty finding good men to assist them. The pay was small, the work was arduous, and the men went on foot, pushing the bleating sheep for months at a time if it was a long, cross-country drive. Of necessity drovers hired outlaws, drunks, drifters, and others who knew nothing of trailing sheep. On the trail they encountered all manner of difficulties. The sheep trail, unlike the cattle trail, which had a Dodge City or a Wichita or an Abilene, had no "sheep towns" full of bordellos, saloons, and gamblers waiting to divide the trailer from his hard-earned wages. Sheep-trailing offered little pleasurable excitement. Many hands deserted after going only a hundred miles or so.

Most drovers, like *pastores*, found dogs indispensable. Some used the large Mexican breed that *pastores* kept as much for protection as for its ability to handle sheep. Others preferred collie-sized Australian shepherds or smaller black-and-white border collies originally imported from Scotland. These types possessed herding instincts that over many generations had been bred into them. Good dogs never nagged the sheep, but simply persuaded them with self-assured leadership. They were obedient to the drover and loyal to the flock. In the Panhandle in 1877, after an outlaw killed his owner—a Navajo boy—and another big sheep dog, a painfully wounded and starving dog continued to herd his flock for a week before cowboys rescued him and the sheep.[1] Some dogs attacked Indians and, shot full of arrows, had to be pinned to the earth with spears and killed with stones. On the trail, dogs kept sheep moving in the right direction, watched for danger, and brought the strung-out sheep to camp at bedding time. "While a sheep is the gentlest of animals," wrote one drover, "when it sets its head to anything it is almost impossible without a good dog to stop it."[2]

Many drovers also used goats. The Spanish goat with its lean, tireless body was the most popular breed. Calling them point goats or bell leaders, trailers put them at the head of the flock, expecting, more often than not with success, that the sheep would follow. They assigned one goat for each 1,000 to 3,000 sheep. The goats led sheep over shaky bridges, through narrow canyons, across shallow streams, and into secure holding pens. Sheepmen often placed bells on the goats. The bells attracted the sheep's attention and allowed the herders to keep track of the animals. *Pastores* trained bell leaders by making household pets of kid goats. Later Anglo sheepmen and experienced trailers adopted the same method. They fed the kid with a bottle, petting and pampering him, and giving him free run in the farmyard. Many of the bell leaders were given names.

Martin H. Kilgore, who came to Texas from Illinois in 1881, found that one of his best herders was an old packhorse. Old Pack, as he was called, learned to get behind sheep on the trail and nose them along. When his sheep got to eating prickly pear, Kilgore with his mates could not budge them. But Old Pack soon caught on that the sheep could be moved when frightened. The pony would stop and give his pack a vigorous shaking. The tin cans and cooking utensils rattling from his sides would stampede the sheep each time. Nobody, Kilgore claimed, got more pleasure out of the little stampede than the horse himself.[3]

[1]Pauline Durrett Robertson and R. L. Robertson, *Panhandle Pilgrimage*, p. 138; John L. McCarty, *Maverick Town, the Site of Old Tascosa*, p. 34; J. Evetts Haley, *Charles Goodnight: Cowman and Plainsman*, pp. 280–90.

[2]Henry S. Randall, *Sheep Husbandry*, pp. 284–86; William Edgar Hughes, *The Journal of a Grandfather*, p. 40.

[3]Florence Fenley, *Old Timers, Their Own Stories*, p. 19.

Equipment for a drive varied, depending upon the habits of the drover and the length of the drive. Commonly it included a light two-horse wagon with a feed trough attached behind, riding horses, a good tent, buffalo robes, blankets, a water bucket, tea kettle, coffee pot, frying pan, small pot, plates, cups, knives, forks, and other small utensils and containers plus food supplies for the men and a little feed for the horses. In addition, most drovers carried some old clothes, a gun, an ax, an auger, plenty of rope, and a few leather strings. Such equipment enabled the drover to camp wherever he could obtain wood and water.[4]

On the long drives one man rode horseback. Called a rustler, he was a sort of foreman. His job was to oversee the drive. The wagon driver was often the cook. Some drovers carried along portable corrals in which to pen the sheep. William Edgar Hughes, who drove sheep into Texas from Illinois in 1860, described his canvas corral as "a very clever and handy device." The canvas was a yard wide with a rope about the thickness of a clothesline running through the top and bottom. About every eight feet along the canvas a well-sharpened, hardwood stake five feet in length was attached to the rope and tent. Drovers sunk the stake into the ground, and the stake held the canvas taut and upright. It made a secure home for sheep. Hughes reported that "wolves in numbers would howl about it; but not one dared to go over it." After the first night or two on the trail, sheep entered Hughes's corral on their own accord.[5]

Considerable money could be made trailing sheep. Richens ("Uncle Dick") Wootton, fur trapper, scout, and Indian fighter, claimed to have made nearly $50,000 in 1852, when he drove 9,000 sheep to the California goldfields. Kit Carson, the famed southwestern scout, made $30,000 the following year, when he trailed some 6,500 sheep to Sacramento. While few drovers made such substantial sums, most of them did enjoy a good return on their investment. At the end of the trail many drovers sheared and sold the fleeces before selling the sheep. The sale of the wool often paid for the cost of the drive, thereby leaving the proceeds from the sale of the sheep as clear profit.[6]

But there were losers, too. In the early days Indians wiped out

[4] Edward N. Wentworth, *America's Sheep Trails*, pp. 73–74.
[5] Hughes, *Journal of a Grandfather*, p. 39.
[6] Ogden Tanner, *The Ranchers*, pp. 89–90.

some flocks on the trail. Mountain lions, wolves, and coyotes attacked at night. Lack of water often depleted flocks before drovers got them to their destination. Sheep drowned trying to swim rivers. Inexperienced trailers sometimes pushed sheep too fast, resulting in loss of weight in the animals and such a poor appearance that selling prices dropped. Rainstorms scattered flocks. Cattlemen sometimes stampeded their herds through a flock, killing and dispersing many sheep. One drover wrote that "my greatest trouble was in keeping other little bunches of sheep, that we met with in the open country, from rushing into our big flock." His route, he noted, was entirely through open country, where there were small bunches of sheep belonging to men with small ranches or settlers who had just moved to the region. When his larger flock came anywhere near them, the little bunches would rush in among his band. "A dozen men on foot or horseback could not keep them out," he said, and "we lost much time in segregating these small bunches from our own flock."[7]

Nevertheless, the possibility of quick returns attracted many drovers. Some, like Wootton and Carson, had never handled sheep before hazarding the trail drive. Experienced or not, sheepmen trailed hundreds of thousands of sheep in the nineteenth century. They moved the animals to all parts of the country. A small number went to Oregon with the first pioneers there. Scores of thousands left New Mexico for California in the gold-rush days of the 1850s. Later tens of thousands trailed out of California for the Rocky Mountain and Plains states. Herders drove sheep to other states, as well as into far-flung regions of Texas. In addition, more localized trailing occurred as drovers took sheep to slaughter, to feeding operations, and to unsettled ranges.

While it waxed and waned in the late nineteenth century, the sheep-trailing industry in Texas was generally an upbeat operation. It started in earnest before the Civil War, picked up after the conflict, and reached a peak in the early 1880s. Several factors affected it—tariff laws, wool prices, weather, western settlement, and new operations in neighboring states, among them. From the end of the Civil War to the 1880s the sheep-trailing industry enjoyed average prosperity. Then in the early 1880s the industry boomed. As scores of western ranchers took up the wool-growing business, there was a big demand for trailers

[7] Hughes, *Journal of a Grandfather*, pp. 40–41.

who were willing to move the exasperating, silly animals to Texas rangeland.

Because of the large demand for sheep, Will and Charles Eaheart during the summer of 1881 took to sheep-trailing. They gathered together four thousand dollars in Shackelford County and forwarded it to a bank in Las Vegas, New Mexico, where they intended to buy sheep to trail back to Texas. After a hazardous trip to Las Vegas, they collected their money and proceeded some fifty miles southwestward to a range of rugged mountains, where some New Mexican *dons* grazed their animals. After a week of sorting and buying sheep, the brothers, with some Hispanic herders they hired to help, started home. They moved eastward through a valley they incorrectly thought was well watered and about twelve to fifteen miles wide. It turned out to be about fifty miles wide and a dry, sandy desert region with little vegetation. The only water they got, they obtained by spreading their wagon sheets and catching the rain. The horses and sheep subsisted on what moisture they could get during the rains and the morning dew. The sheep tended to spread out over the wide valley, grazing leisurely on the sparse vegetation. The drive took many weeks longer than expected. Indeed, as summer turned to fall, the brothers decided to drift their animals down the Pecos to Grand Falls rather than across the dry tableland of the Llano Estacado in eastern New Mexico and western Texas. At Grand Falls they waded their sheep across the river and took them to the Colorado City area, where they arrived in the middle of winter. While Charles remained with the sheep there, Will returned to New Mexico for another flock.

This time Will Eaheart rode to Socorro, just across the Rio Grande in Socorro County. When he found a man willing to sell an unusually good flock of American Merinos, he wired Charles to come out as quickly as possible to help him on the return trip. After closing the deal, Will joined his brother near Socorro. They started the sheep home. When they drove the Merinos across the Rio Grande, 800 head drowned. Afterward a mountain lion followed them, making nightly raids on the flock. At White Oaks in the Jicarilla Mountains of Lincoln County they came to the legendary Malpi Mountain, one of the unusual formations of the West. Because of the numerous caves, great boulders, and bottomless pits, the mountain was nearly impassable for sheep. The pass through the mountain was a toll road, but the owner wanted several hundred dollars to let the sheep through. The brothers, rather than pay the heavy cost, chose to risk a few sheep to the pits and find their own way. After a hard week of slow but steady progress they made it with the loss of only a few head. From there they drove their animals without further incident to Colorado City, mixed them with the original band, and pushed the combined flocks, numbering about 8,000 head, to the vicinity of Abilene, reaching there in the spring of 1882.

A few nights after setting up camp next to the mouth of a ravine near Abilene, the brothers had more trouble. A sudden cloudburst of rain and a tornado swept through the camp. The wind carried away their big tent and scattered supplies over a wide area. The downpour caused water to come rushing through a ravine. It carried away most of the rest of the camp equipage, scattered the sheep, and forced the men to scramble for cover to high ground. When he reached safety, Will could see nothing left of the camp, and the sheep he saw were floundering in water. By morning, when the weather had cleared, the brothers found that they had lost 200 animals.

Discouraged, but not broken, the men gathered their sheep and drove them to dry land. Despite the hazards and misfortunes, they sold the sheep at a good profit. For three more years the Eaheart brothers with their herders made annual trips to New Mexico to buy sheep and trail them back to Texas. They quit only after their 1885 trip, and by then the Texas sheep boom had collapsed.[8]

Some people, among them Harry Holdsworth, got their start as wool growers by trailing sheep. In 1882 Holdsworth and his father agreed to drive 1,200 head of sheep from Benavides in Duval County to San Antonio, a distance of 180 miles, to be marketed. They trailed the animals with a dog and a lead goat, carrying their supplies on two pack mules. They made about 12 miles a day, a little above average for trailing sheep. The sheep started each morning at a good pace. At noon they halted to rest under the shade of trees or shrubs, or in the absence of shade crowded together to hide their heads under one another's flanks. While the sheep rested, the men caught a welcome nap. In the late afternoon the flock moved leisurely out again along the trail.

[8]Elizabeth Ann Maxwell, "Experiences of the Eaheart Brothers, Sheep Traders," *West Texas Historical Association Year Book* 30 (1954): 157–61.

In the evening the Holdsworths with their dog brought the animals into a bed ground where water and good grass were available. Here they prepared their main meal for the day. Afterward they cleaned the cooking utensils and relaxed, sipping hot coffee, smoking their pipes, and recounting the day's long drive. This routine they kept up for the sixteen days they were on the trail. As pay they received 100 sheep. These they drove without difficulty to Zavala County and added them to 1,000 others they had acquired earlier in the same manner.[9] Holdsworth repeated this kind of sheep-trailing operation until he had a large enough flock to begin ranching on the Edwards Plateau.

Although Holdsworth had little difficulty, Arthur G. Anderson was not so fortunate. In the early 1880s Anderson, pleased with the Rambouillets he had acquired in California a few years before, made a second trip to that state. There he bought a band of several hundred head. By this time the Southern Pacific Railroad had been completed as far east as Texas' upper Big Bend country, enabling him to ship his new stock there. However, when he tried to trail his sheep overland across the section, cattlemen stopped him. They demanded that he comply with a Texas law that forbade the driving of a band of sheep numbering more than twenty-five head onto land not belonging to the sheepman when the landowner objected. Refusing to comply with the law, Anderson was arrested and, because the county had no jailhouse, placed in a dry cistern under the Fort Davis courthouse floor. Each day authorities took him out and ordered him to pay a fine of $100. Each day he refused and reentered the cistern. At the end of a week, when he agreed to give bond of $500 to appear for trial later, county officials released him to round up his sheep, which had been thoroughly scattered by local cowboys. Anderson, with the help of herders he had hired to drive the sheep, rustled up most of the animals and moved to his ranch in Mitchell County. He forfeited the bond.[10]

When he returned to his ranch, Anderson found Texas wool growers suffering from low prices they blamed on the tariff of 1883. The new tariff left the specific rates on clothing and combing wool as

[9] Fenley, *Old Timers*, pp. 1–16.

[10] "Blazing the Way," *Sheep and Goat Raisers' Magazine* 7 (December, 1926): 4; Arthur G. Harral, "Arthur G. Anderson, Pioneer Sheep Breeder of West Texas," *Southwestern Sheep and Goat Raiser* 6 (November, 1935): 12–13, 31; newspaper clipping, Morgue Files, Texas Sheep and Goat Raisers Association, Southwest Collection, Texas Tech University, Lubbock.

they had been (except for a slight change in the dividing line in respect to value), but it eliminated the ad valorem rates. Texas wool had been selling for twenty-four cents per pound. Since the ad valorem charge applied only to wools valued at more than thirty cents per pound, the law of 1883 did not represent a significant change for Texas wool.[11] Yet, Texas wool growers got as little as six to seven cents per pound for wool in 1883. The tariff, they believed, was the cause. Many sheepmen, such as Joseph Tweedy, Lawrence Grinnell, and J. B. Reynolds, went broke. Others, including William L. Black, suffered great financial losses. Sheep-trailing, although affected, continued to be profitable for a couple of years longer.

Meanwhile, wool schedules became the most important and most sharply debated part of the tariff system. Each year Texas wool growers pressed for revision of the 1883 law. Not until 1890 were they successful (see table 6) In the McKinley Act, passed that year, Congress retained the division of wool into three classes: clothing, combing, and carpet wool (see appendix). The changes in duty were in the main significant for their direction rather than for their amount. Indeed, Congress slightly raised the duties on clothing and combing wool to put them where they had been before 1883 and to placate growers who blamed the drop in wool prices on the reduction of the protection rates that year. On carpet wools Congress adopted a more radical change: ad valorem rates of 32 percent for prices up to thirteen cents and 50 percent on higher prices. But probably little carpet wool was produced in Texas.[12]

The Wilson-Gorman tariff of 1894 was more significant. A revolutionary measure, it allowed wool to enter the United States duty free. In response, foreign wool growers dumped huge inventories on the American market. The entire Texas sheep industry was adversely af-

[11] Frank W. Taussig, *The Tariff History of the United States*, p. 239; Chester W. Wright, *Wool Growing and the Tariff: A Study in the Economic History of the United States*, p. 271.

[12] Taussig, *Tariff History*, pp. 256–58. There is some difference of opinion on how much carpet wool came from Texas. The carpet wool manufacturers insisted that very little, or no, carpet wool was produced in the United States in the nineteenth century. They were trying to justify much lower duties on carpet wools than on the other wools. The chief characteristic of carpet wool was its coarseness, and during the nineteenth century there was a fair amount of coarse wool raised in the United States, including Texas. The wool growers suspected that a good deal of wool suitable for clothing was coming into the country as "carpet wool" and thus paying a low rate.

TABLE 6
Tariffs and Wool Duties, 1867–1929

Act of	Class I Clothing Wool	Class II Combing Wool	Class III Carpet Wool
1867	Value up to 32¢— 10¢ per lb., plus 11% ad valorem Value over 32¢— 12¢ per lb., plus 10% ad valorem	Same as duties on Class I	Value up to 12¢— 3¢ per lb. Value over 12¢— 6¢ per lb.
1883	Value up to 30¢— 10¢ per lb. Value over 30¢— 12¢ per lb.	Same as duties on Class I	Value up to 12¢— 2½¢ per lb. Value over 12¢— 5¢ per lb.
1890	11¢ per lb.	12¢ per lb.	Value up to 13¢— 32% ad valorem Value over 13¢— 50% ad valorem
1894	Free	Free	Free
1897	11¢ per lb.	12¢ per lb.	Value up to 12¢— 4¢ per lb. Value over 12¢— 7¢ per lb.
1909	Same as 1897	Same as 1897	Same as 1897
1913	Free	Free	Free
1922	31¢ per lb.	31¢ per lb.	Free for making carpets. If not, 12¢ per lb. Penalty 20¢ per lb. if not used for carpets
1929	31¢ per lb.	31¢ per lb.	Free for making carpets. If not, 24¢ per lb. Penalty 50¢ per lb. if not used for carpets

SOURCE: Karl Everett Ashburn, "Tariffs and Wool Duties since 1867," *Sheep and Goat Raisers' Magazine* 10 (November, 1929): 104–106.

fected. Since wool for years had been a key part of the protective system, the 1894 tariff had important political and economic implications extending beyond the sheep industry. Republican Congressmen had supported all-inclusive protection since the Civil War, and through the wool duties Congress had provided the protection Texas agricultural interests wanted. There were duties on most major agricultural products produced in the United States, such as wheat, corn, and meats, but these products were continuously exported and less affected by import taxes. Wool, however, was imported. Thus, in the 1890s growers found themselves competing at such a price disadvantage that they lost money and abandoned sheep raising. In Texas, when wool prices dropped in the mid-1890s, sheep trailers found their area of employment restricted. Many of them turned to goat and mohair production, accepted the settled life of herding sheep—when they could find such work—or quit agriculture for industrial employment. After 1894 Texas sheepmen, finding fewer herders, encountered labor shortages that have continued to plague the industry to the present time. While there were sheep drives afterward, the sheep-trailing industry never regained its importance. How much of all this can be attributed to the Wilson-Gorman tariff with its duty-free wool is hard to determine, but wool growers in Texas and elsewhere blamed their economic woes on the absence of tariff protection.

Along with tariff duties, weather also affected the sheep trailing industry. Drought in the mid-1880s hurt many Texas wool growers who had overstocked their sparse range. Conditions became so deplorable for sheepmen that many of them left the industry and the range country of the West. Some counties in the Edwards Plateau saw their populations decline as ranchers left to return to more settled eastern portions of the state and to jobs or farms they had left behind. Much the same thing happened again in the early 1890s.

Even more dramatic was the long, cold winter of 1886–87. It was complicated by drought conditions that had prevailed in the summer and fall. The severe weather began when a great blizzard hit Texas on New Year's Eve, 1886. Christmas week had been clear and pleasant until late on the last day of the year when a huge cloud bank reared up to the north. Ranch folks scurried about to get set for a norther. They penned some sheep and cattle and hurried to drive others to protected spots on the range. Without other warning, the blizzard struck, driving a roaring sheet of dry sleet before it. Visibility was limited to a few feet. By daybreak on New Year's Day, 1887, the wind had become a tearing, grinding monster that neither man nor animal could face. The icy blast pushed frightened, stumbling sheep and cattle southward until the animals blindly bumped into some of the new barbed-wire fences that had been strung along property lines. There they died, trampled or suffocated. As the snow continued to whirl and blow, the first waves of livestock pressed on against the fences and piled up to form bridges over which succeeding animals crossed to continue their flight.

Isolated ranch families and their help also took a terrible beating. Some settlers, living in flimsy shacks that could not withstand the wind, perished. Those in substantial houses and those living in dugouts fared better. A few herders died in their saddles. Some travelers in wagons froze to death along with their teams. It was a grim tale. After a few days even those people in adequate shelter began to suffer for lack of fuel or food. Families huddled under bedding for days, lacking heat and nourishment. They burned scrap material, boxes, and furniture to keep from freezing.

Early on the morning of the tenth day the blizzard blew out. The sleet stopped, the skies cleared, the sun shone. Ranchers emerged from their shelters to survey the damage. Riders hastened to town or to large ranch headquarters for food and medical supplies. Others gathered timber, brush, or cow chips to replenish fires. After tending to the personal needs, the stockmen went in search of their sheep and cattle. They found only a few alive. The dead cattle they uncovered more often bore the brands of neighbors living far to the north. The sheep they stumbled across in semiprotected draws and ravines were buried under snow and packed so tightly together that they had smothered. A few ranchers set about to salvage cowhides, but with little heart for the job they soon quit. Instead, after recording brands on the dead cattle, they burned the animals. Packs of wolves and coyotes and flocks of buzzards ravaged the corpses.

Late in the month another blizzard howled across the Edwards Plateau. On the heels of this storm came a numbing cold that drove temperatures below zero. Ranchers and herders huddled about their shelters, unable to do more than worry about their livestock on the range. In the spring, when they finally could go in search of their animals, writes one chronicler, they "saw a sight they spent the rest of their lives trying to forget." They found carcasses heaped upon one another in every ravine, gaunt animals staggering along on frozen feet, and piles of dead bodies along the fences. The winter had brought death to several people, broken health to others, and financial ruin to many.[13]

Despite such hazards, the sheepmen and drovers persevered. Martin Kilgore was one of them. In September, 1886, he moved five thousand sheep on to rangeland in Menard County. Recent rains after a dry summer had brought out the grass and weeds. It was, he thought, an ideal place to winter his flock. But over the next two months no more rain fell. The fall turned dry and hot, and winds sapped the last moisture from the ground. What vegetation remained turned brown, curled up, and blew away. Water holes dried up to cracked acres of baked earth. "I had to skin out again hunting grass," Kilgore said.[14]

Having only an old one-eyed Mexican herder, he hired three outlaws to help him with the sheep. The first night out the *pastore* stole Kilgore's rifle and vanished. Kilgore and the others pushed ahead with horses, wagons to carry water for themselves and the horses, a dog, and the sheep. They crossed the Edwards Plateau tableland pushing westward, going twenty-one days without finding water for the sheep. Farther west, however, the grass was green and dew heavy enough so that the sheep made out, although crazy for water. After the water barrels emptied, Kilgore went ahead to find the precious resource. "I knew," he reported, "that the dews [in the hills] were so heavy that enough moisture was deposited on solid rocks to form small puddles of water of mornings." He went up to a knoll and found enough to collect two or three gallons. After returning with the emergency water, he set out to locate Beaver Lake. Finding it, he returned with a sufficient water supply in barrels for the men and horses. The sheep were still subsisting on morning dew. Deciding to drive the sheep to Beaver Lake, Kilgore abandoned the wagons temporarily, lined his men up on saddle

[13] Ray Allen Billington, *Westward Expansion*, p. 596. For the descriptions of the blizzards, see also newspaper clippings, Morgue Files, Texas Sheep and Goat Raisers Association, Records, Southwest Collection, Texas Tech; David J. Murrah, *C. C. Slaughter: Rancher, Banker, Baptist*, pp. 57–58, 150; and Frederick W. Rathjen, *The Texas Panhandle Frontier*, pp. 242, 244–45.

[14] Fenley, *Old Timers*, p. 19.

horses, and placed himself in front of the sheep. He stayed in the lead of the sheep to head them off and hold them down, for he knew that when they smelled the water, they would stampede. The next day they reached Beaver Lake. It covered an acre or two, but, said Kilgore: "When those sheep ran into it, you couldn't see any water at all. They just covered it. They drank till first one leg then another would raise up because their paunches were just like drums." Although afraid his animals might die, Kilgore did not lose a sheep. That night when he and his outlaw herders butchered an animal for fresh mutton, the men found a full gallon of water around its intestines.[15]

After resting for a couple of days, the party retrieved the wagons and pushed on to the Pecos River, reaching it in December at a place called Tardy Crossing. Here they camped through cold winds and snow, seeking the best place they could for protection. The river presented another problem. It was about one hundred feet across at that point, two to three feet deep with swiftly running water. Hearing that a new building was going up on a ranch across the Pecos, Kilgore at dusk hitched a wagon, waded the river, and started for the construction site. Winding down a steep slope on the opposite bank of the river, he lost a wheel on the wagon. It went flying down the hill ahead of him and rolled into a camp of Mexican grass cutters. The Hispanics were asleep at the time, Kilgore noted, "but when that tire rolled in amongst them there was a scatteration." Kilgore explained what had happened to the surprised workers. They gave him some old boot leather, and with it he remounted his wheel and continued. At the building site Kilgore purchased some thirty pieces of lumber before returning to Tardy Crossing.

Back at the Pecos, Kilgore used the lumber and the wagons to bridge the swift stream. He spaced the three wagons across the river, connecting them with the lumber he had purchased and weighing them down with rocks. One of the men took a pan of salt and started across the makeshift bridge, coaxing the sheep after him. Hungry for salt, the sheep needed little prodding. They followed without difficulty. "In an hour's time," said Kilgore, "we had crossed the whole flock, and then we moved the camp across. Everything was so happy that our old shepherd dog presented the camp with pups that night."

[15] Ibid., pp. 19–20.

Kilgore moved on to range country below Dryden in Terrell County, paid off his outlaws, and hired two capable Mexican herders. With only four sections of fine grazing land for his five thousand sheep, but plenty of water and capable help, Kilgore made out. His ewes produced a 96 percent lamb crop, and his sheep became fat.[16]

But Kilgore found that his sheep-trailing troubles were not over. In the spring he sold his animals to the J. M. Shannon Company, agreeing to hold the ewes and lambs until they could travel. Agents for the company removed the others to Paisano Pass along the railroad near Alpine. When the remainder were ready, Kilgore got a map or sketch of the route to Paisano Pass and started out. Again he took a wagon for supplies, a dog, and extra help. A young tenderfoot to sheep-trailing drove the wagon. Kilgore followed the sketch to a designated windmill, where he got additional directions from some ranch hands. The cowboys indicated that if he went straight across the country rather than following the road, he could cut off about twelve miles of his trip. He told the young wagon driver to go around by following the road, a fifteen-mile trip. Kilgore and the herders, meanwhile, would drive the sheep straight across through some rough gullies and ravines and connect with the road three miles distant. The tenderfoot drove beyond the meeting point and continued until he hit the railroad. There he abandoned the sheep drive, the sheep, and his companions and took the train eastward. The station agent found the wagon and team and cared for the animals until Kilgore could claim them.

"Well," noted Kilgore later, "those two herders and I started out across the mountain afoot and we spent three of the hottest days I ever saw in my life." About halfway over the sheep refused to move. They waited for the boy to bring water, but of course, he never came. At midafternoon on the second day, the two herders went crazy, stripped off their clothes, and lit out for water. Kilgore remained with the sheep until dark. Then he drove them into a secluded parapet in Chalk Valley before heading back to the windmill. He was several hours getting back to the place. Half-crazy himself, he walked one or two hundred yards—thinking it was a great distance—and fell down to sleep. At the windmill he drank slowly and carefully, taking an hour or more to drink what he wanted. At daylight he walked to a ranch about seven miles

[16] Ibid., pp. 20–21.

away to get help. With three men and horses he went back to the sheep and rounded them up. By this time it was noon. The men prepared a light lunch and ate it. But then one of the men threw a match or cigarette down and set the range grass afire. They fought the blaze with saddle blankets. Not until they were exhausted, smutty, and burned did they succeed in getting the fire out. Afterward they took the sheep to water. Kilgore then reassembled his wagon outfit and took the sheep on to Paisano Pass without further incident.[17] If not easy, sheep-trailing was at least interesting for Martin Kilgore.

James Prentice, one of the first men to run Angora goats in West Texas, remembered sheep-trailing, even in the mid-1880s as a more pleasant experience than Kilgore's. Before he began raising goats, Prentice in the 1880s herded sheep for D. Q. McCarty in Tom Green County. Then throwing in his lot with a young Bavarian named Charley Mueller, he roamed western Texas for several years. With a wagon, team, and tent, they hired on as sheep trailers, moving sheep from county to county on the Edwards Plateau for ranchers who had just purchased the animals. "Those were happy days," said Prentice, "plenty of game, no hunting restrictions, and all the time we wanted. Nobody ever got in a hurry." The low prices and bad weather of the mid-decade bothered them only a little, for they always tried to get part of their pay in sheep. Over the years they built a sizeable flock. But drought and the national economic depression of the 1890s turned Prentice to Angora goat raising, just beginning to attract popularity in Texas.[18]

One of the last and most difficult long sheep drives into the state occurred in the late 1890s. From Los Angeles, California, Robert Maudslay moved some 4,000 sheep by rail into western Arizona Territory, where he acquired an additional 6,500 head. He was to drive the entire band, numbering more than 10,000 sheep, overland into New Mexico and thence to the Devils River in Texas. He carried his supplies in two wagons drawn by mules. He had ten herders and two drivers, who also did the cooking, to help look after the sheep. His men were a mixed bunch: five Mexicans, two Frenchmen, one American, and one Basque. Maudslay himself was an Englishman. He used

[17] Ibid., pp. 21–22.

[18] Newspaper clipping, Morgue Files, Texas Sheep and Goat Raisers Association, Records, Southwest Collection, Texas Tech University.

neither dogs nor point goats. He kept his men afoot but often used a horse himself. As he traveled across Arizona, his flock increased to 12,000 head.

The drive went well until it neared Winslow, Arizona. There a smallpox epidemic raged out of control. Conditions were so bad that authorities used railroad boxcars as hospitals and isolation centers. Maudslay's entire crew quit and headed back to California. The rustler then hired a motley lot to take their place: a gambler, a jailbird, an alcoholic, and others who, said Maudslay, "were pretty much of the same kidney."[19] They were poor herders, and Maudslay lost many sheep.

The party entered New Mexico near present Red Hill. Here, among the canyons and piney woods of the Mogollon Mountains, its difficulties multiplied. Water could be found only in the gorges, some three hundred feet or more deep, with steep sides. A few had trails blasted out for stock to get down to the water. Others were impassable. Trying to haul a wagon loaded with filled waterkegs from the canyon floor to the rim, Maudslay lost the vehicle. It parted from the mules, flew backward down the canyon trail, flipped over, and splintered into a mass of broken timber and twisted iron. To replace the irreparable wagon, Maudslay had to buy another at an exaggerated price. The thick woods and deep canyons allowed sheep to separate from the main flock. Thus, during the day Maudslay spent much of his time hunting lost animals or finding a route to take them through the rough country. Since the gambler refused to go and Maudslay did not trust any of the others, he occasionally made a trip to a nearby town for fresh supplies. Each time he returned, he found more sheep gone. "I don't remember," he said, "how long I rambled backwards and forwards through the pine forests trying to get [the sheep through]. Night after night I helped the herders bunch their sheep on the bedding ground, and then, leaving one man on guard, gathered with the rest around the cook's fire to eat beans, bacon, dried fruit, and camp bread that was the welcomed supper."[20]

Out of the mountain roughness Maudslay and his men took their

[19] Cited in Winifred Kupper, "Sheep Drive in the 90's," *Sheep and Goat Raiser* 26 (December, 1945): 41.

[20] Ibid., p. 43.

sheep. They drove them through brush, tableland, and prairie. Through desert stretches they sometimes moved at night, holding the sheep in uneasy rest during the heat of the day, allowing them to graze in late evening and early morning, and taking them to water when they found it. They passed into Texas east of El Paso, cut southeastward through the upper Big Bend country, and followed, at a distance of several miles, the general course of the Texas Pacific. At the Pecos they crossed the sheep to the east side. The wagons, following a faintly defined trail downstream, crossed and recrossed several times. Carelessness on the part of one of the drivers caused a wagon to get stuck in the middle of the river. Maudslay saw that there was nothing to do but unload it and carry the contents to the brushy bank. Next they took the wagon bed and floated it out. Then they uncoupled the big wagon wheels and rolled them out two at a time. "It was a long, back-straining task," remembered Maudslay, "in which I had plenty of time to reflect that not all incompetence and stupidity of hired help on a sheep drive lay in the herders."[21] After that difficulty they reached the Devils River without further incident.

The trailing of goats never assumed such proportions as sheep-trailing. It did not differ markedly from sheep driving. Goats could travel faster, moved in smaller bands, and needed to forage on green shoots of trees and shrubs, but otherwise there was little difference. Indeed, in the Spanish period, goats moved with sheep. While some trailing occurred in the mid-nineteenth century, major movement of goats did not take place until the 1890s and afterward, when the Angora goat and mohair industry gained a foothold in Texas. By then railroads, fences, and herding laws made a long overland goat-trailing industry unnecessary or impossible. But irregular, shorter-range goat-trailing centered in the Hill Country and in the southwestern portions of the Edwards Plateau.

A drive occurring in the twentieth century revealed some of the hazards of goat-trailing. Fred Gipson, author of the frontier classic *Old Yeller*, was one of three herders who moved 700 mixed goats from a ranch south of Mason to some hilly rangeland in Blanco County about ninety miles away. The herders strung out the goats, with one man on horseback riding point, another on foot in the rear, and the third, a

[21] Ibid., p. 50.

youth, inexperienced in the goat business, driving a Model T Ford loaded with bedrolls, water, and supplies. They moved out slowly at noon late in October, settling and trail breaking the goats. The first afternoon they made seven miles without difficulty, putting the animals in a local rancher's pens before sunset. Trouble started the next afternoon. A sudden, heavy rain struck while the herders moved their goats along a creek bed. In a few minutes water came rushing down the creek, and in thirty minutes the goats stood shivering and bleating under every bush for a mile up and down the creek. When the Model T stalled, the men had to wade waist deep into the cold water to get the vehicle out of a draw before rushing water swept it away. The goats, refusing to move from their cover, stood under the brush all night. By morning the rain had stopped, but it had left the trail mud-soaked. The stubborn goats had to be shoved along. The herders, working eighteen hours a day, pushed the wet, belligerent animals through mud and water. For three days the beleaguered herders urged on the goats. They slept in cold and wet blankets, lost their food supplies, and struggled to get the battered Model T across muddy creeks. When they moved into live-oak country northeast of Fredericksburg, the rain fell again. A couple of days later they herded the goats into a large pasture, and the trail animals got mixed with the local rancher's prized goats. The tired herders spent nearly a full day cutting out their band and getting back on the trail. Two days later, hungry, irritable, exhausted, they delivered the trail-weary goats at the destination point in Blanco County.[22]

To the end of the nineteenth century sheep-trailing represented an important phase of the range sheep industry in Texas. Tariff laws, weather conditions, predators, and low wool prices affected the business, but it remained a significant part of Texas livestock production until well into the 1890s. The excellent opportunities for quick profit encouraged all manner of men to hazard the difficult work and lonely months of pushing defenseless animals across wild country. In the twentieth century there were occasional long sheep drives, but the spread of railroads and settlement soon brought the sheep-trailing industry to an end. By that time, the Trans-Pecos country had become an important center for Texas sheep raising.

[22]See Fred Gipson, "Easy Money This Goat Money," *Sheep and Goat Raiser* 50 (October, 1969): 51, 59.

10

The Trans-Pecos

SHEEP and goat production in the vast Trans-Pecos region began with early Spanish missionaries and civilian settlers. It all but disappeared in the early nineteenth century, but started anew with Milton Favor and others about the time of the American Civil War. It boomed in the 1880s, then leveled off or even retreated in the following decade before reestablishing itself after the turn of the century. In many ways the range sheep industry in the Trans-Pecos was an extension of ranching activity from the Edwards Plateau. Goat production did not reach impressive proportions there until after the open range had disappeared.

The Trans-Pecos is an empty and isolated area of Texas stretching westward from the Pecos River and northward from the Rio Grande to the New Mexico border along the thirty-second parallel. Although composed of only nine counties and portions of another, it covers more than 28,000 square miles. The land is semiarid, rugged, and sparsely settled. An agricultural region, it is varied in its topography, healthful in its climate, and immense in its extent. Traversed from north to south by an eastern range of the Rocky Mountains, it contains all of Texas' true mountains.

Physiographically the Trans-Pecos is divided into several parts. The eastern third of the huge area is a rolling to rough country lying in the valley of the Pecos River and on the Stockton Plateau at the eastern base of the Davis Mountains. With only ten to twelve inches of rainfall annually, this was exclusively ranching country until long after the end of the open range, when the discovery of large quantities of ground water brought irrigation for cotton, alfalfa, and other crops.

In the northwest are the Guadalupe Mountains, the highest in the Trans-Pecos, and the Diablo basin. The United States government maintains Guadalupe Mountain National Park in the region just below the New Mexico border in Culberson and Hudspeth counties. West of the Guadalupe range lies the Diablo basin. Having no drainage outlet to the area, runoff from the scant rain that falls there accumulates in a series of salt lakes. The lakes, dry during periods of low rainfall, were for many years a source of commercial salt. Few sheep graze there today.

The Davis Mountains, a third feature, are principally in Jeff Davis County. The highest peak, Mount Livermore at 8,382 feet, is one of the highest in Texas. These mountains interrupt the moisture-bearing winds and cause more precipitation than elsewhere in the Trans-Pecos. They contain more grass and trees than other regional mountains.

South of the Davis Mountains lies the Big Bend country. The Big Bend is a mountain area of scant rainfall and sparse population. Along the Rio Grande, which encompasses the Big Bend on three sides, are Santa Elena, Mariscal, and Boquillas canyons, each reaching a depth of nearly two thousand feet. Because of the remarkable topography and plant and animal life, the federal government has set aside the extreme southern part of this region as the Big Bend National Park.

The final part, the Upper Rio Grande Valley, is a narrow strip of irrigated land running down the river from El Paso for a distance of seventy-five miles or more. In this area are the historic towns and missions of Ysleta, Socorro, and San Elizario, among the oldest in Texas. Cotton is the chief product of the valley. This limited area has a dense population, in marked contrast to the country surrounding it.

With as little as eight inches of average annual rainfall, long hot summers, and usually cloudless skies, the Trans-Pecos area produces without irrigation only drought-resistant vegetation. Grass is usually short and sparse. The principal growth consists of such shrubs as lechuguilla, ocotillo, yucca, cenizo, and other arid plants. In the non-arid regions yeso, chino, and tobosa grasses prevail. There is some mesquite. The grass, especially on the higher mountain slopes, includes many southwestern and Rocky Mountain species not present elsewhere in Texas. On the desert flats, black grama, burrograss and fluffgrass are frequent. Less arid, more productive, sites have numerous species of grama, muhly, Arizona cottontop, dropseed, and perennial three-awn grasses. At the higher elevations, plains bristlegrass, little bluestem, Texas bluestem, side-oats grama, chino grama, blue grama, piñon ricegrass, wolftail, and several species of needlegrass are common.[1]

For livestock raising the Trans-Pecos is an excellent region. Today ranchers graze about 500,000 head of cattle west of the Pecos and an estimated 300,000 sheep. Perhaps 80,000 goats browse the low shrubs and semiarid grasses. Raising livestock in the territory, however, is not without its difficulties. Ranchers lament the low wool and mohair prices, the expensive transportation costs from the remote area, and the ineradicable predators that, they claim, kill their flocks.

The modern-day complaints mirror the grumbling of Anglo stockmen a hundred years ago. In the late nineteenth century, as one Texas sheepman arrived in the Trans-Pecos with his flock, he groused, "When I put those three hundred ewes on the place, there was a coyote for every one of 'em, and a bobcat trailing every coyote."[2] He could have noted that mountain lions, wolves, buzzards, eagles, and Indians also disturbed the sheep. But the first sheepmen there found empty land, plenty of good grass, available water, and a high upland country comparatively free of livestock diseases.

Sheep had entered the upper Rio Grande Valley portions of the Trans-Pecos in 1598 with Juan de Oñate and other Spaniards. In the seventeenth century as additional colonists filtered up the river to New Mexico, Spanish farmers, soldiers, and missionaries trailed sheep northward to stock their ranches. Sheep multiplied rapidly in the upper Rio Grande, and within a couple of dozen years herders moved thousands of them southward to Chihuahua and Viscaya.

Sheep and goats also formed part of the flocks that supplied missions in present Texas at Ysleta, Socorro, and San Elizario. At civilian communities that grew up near the mission stations, farmers and ranchers maintained sizeable flocks. For nearly two hundred years the timid creatures provided clothing and a stable food supply. Missionaries with their Indian neophytes and civilian stockmen tended the sheep along the river from its junction with the Concho across from present Presidio to above El Paso. Compared with those in New Mex-

[1] *Texas Almanac*, 1980–81, pp. 102–103, 115.

[2] Cited in Winifred Kupper, *The Golden Hoof: The Story of the Sheep of the Southwest*, p. 118. See also Virginia Madison, *The Big Bend Country of Texas*, p. 147.

ico, flocks in the Trans-Pecos were small, never more than five thousand head. There is little evidence that the Spanish held large permanent flocks elsewhere in the region. *Pastores*, however, using the ancient *transhumante* system, followed grazing circuits from the Rio Grande in New Mexico to penetrate deep into the Trans-Pecos highlands with their sheep each summer, before returning to home pastures near the river as winter neared. In the eighteenth century Spanish sheep raising in the Trans-Pecos declined, and after the decision in the 1760s to abandon much of the far northern frontier it all but disappeared. Nearly a century passed before sheep and goat production on a grand scale returned to the region.

From the Spanish decision in 1767 to retreat from most of Texas and New Mexico until the American takeover at the close of the War with Mexico in 1848, the number of sheep in the Trans-Pecos declined. Many Spaniards pulled out, taking flocks with them, and, after Mexico's independence was won in 1821, few Mexicans wanted to carry livestock into the northern border region. At the end of the Mexican War, explorers, soldiers, traders, and others entered the Trans-Pecos, but few stockmen came. After the discovery of gold in the West, thousands of would-be miners crossed portions of the region on their way to California. And in Washington, D.C., Congressmen debated the question of a mail route and eventual railroad to the Far West. For these reasons and others, Anglo-Americans developed an interest in the isolated region. It was not long before they carried in sheep and goats.

Captain John Pope, United States Army, in 1855 supervised one of the first Anglo-American sheep drives west of the Pecos. Captain Pope with seventy-five troops, sixteen head of cattle, and seventy sheep marched overland from El Paso eastward. He intended to explore and survey the country for a proposed rail line. He planned to use the sheep for food, but wolves stampeded them and ate thirty-two head after the party crossed the Pecos River. Nevertheless, Pope completed his task, even stopping to drill a well on the High Plains near where the Texas–New Mexico boundary crossed the Pecos River. The well showed plenty of water, but at a depth too great to be practical.

The Big Bend's first sheepman of note was the cattle king Milton Favor. In 1858 this large man with a flowing beard and iron will, who suffered from tuberculosis, established a headquarters ranch on Cibolo

Creek near present Shafter. An adobe fortress designed for defense against Indians, Cibolo was a rectangular structure with walls twenty-five feet high pierced with loopholes for rifles and thick enough at the top for men to walk. Eventually Favor hired a large crew of Mexican laborers, *vaqueros*, and herders and established four other strongholds. He purchased livestock in Mexico and drove the animals northward across the Rio Grande to stock his ranches. From an early date Favor ran sheep on the same semiarid grassland that he used for cattle. For many years afterward his influence and example of combined cattle-sheep operations pervaded the stock industry in the Big Bend.[3]

Yet in the first couple of decades after Favor built his outpost and ranch in the Big Bend only a few other herders brought sheep to the Trans-Pecos. It was cattle country. Some men, like Robert Casey, who in 1868 rounded up his sheep in Menard County, entered the empty territory only to cross it to New Mexico. Casey, after placing his household goods and family in covered wagons, headed for his new home. When he neared the Pecos, he joined forces with a cow outfit. Indians attacked the combined party on the Pecos and scattered their livestock. A band of warriors went after the sheep, getting off their horses to drive the animals before them. When she saw what was happening, Mrs. Casey grabbed a tin pan and ran out some distance from the wagons calling the animals. When they heard the familiar sound, which to them meant a generous supply of shelled corn, the sheep turned upon their Indian herders and, upsetting every brave who attempted to stop them, ran quickly back to their mistress. Had Casey not rushed to his wife's rescue, Indians would have taken her captive. After the incident the Caseys followed the Pecos without further incident to New Mexico.[4]

Most Anglo sheepmen remained in the east, pushing their flocks up the Guadalupe or onto the Rio Grande Plain. Not until the sheep boom struck Texas did herders enter the Trans-Pecos. Then they overran what are today Val Verde and Terrell counties, physiographically

[3]Robert M. Utley, "The Range Cattle Industry in the Big Bend of Texas," *Southwestern Historical Quarterly* 69 (April, 1966): 423–24; Madison, *Big Bend Country*, pp. 147–48; John Ernest Gregg, "The History of Presidio County" (Master's thesis, University of Texas, Austin, 1933), p. 62.

[4]Roy D. Holt, "Pioneering Sheepmen," *Sheep and Goat Raiser* 26 (December, 1945): 16.

part of the Stockton Plateau, before moving into present Pecos, Brewster, and other counties of far West Texas. When they came, the herders brought large flocks, trailing 1,500 to 2,000 head each. Quickly sheep outnumbered cattle. By 1880 there were not less than 39,000 head, more than double the number of range cattle there.[5]

According to the United States census of 1880, the total population of the Trans-Pecos country was 8,525 people, most of whom lived near El Paso. Some 3,475 citizens were native Texans, and 485 were from the other southern states, including the border states, Kentucky and Missouri. Of the total, 603 residents were blacks, most of whom served in the United States Army. Among the foreign-born inhabitants, there were 3,900 Mexicans and more than 140 Europeans, of whom the greater number were from Germany, England, and Ireland.[6]

A population boom of sorts began in 1881. Not limited to the Trans-Pecos, it included all of West Texas and lasted until at least 1885. During the period and for a number of years afterward, small towns sprang up along the tracks of the Southern Pacific and Texas and Pacific railroads. In the beginning, most of the communities appeared as little more than railway stations, whose importance as shipping points or whose fertile soil and desirable locations attracted settlers. Alpine and Marfa, especially the latter, stand as good examples of communities with favorable sites. In 1883 Alpine, known as Murphysville until its incorporation some four or five years later, contained seven lumber shacks—residences—one general store, and two saloons, each with a dancing-hall apartment. In 1887 it became the county seat of the newly organized Brewster County. Judge J. M. Dean, a pioneer lawyer, in 1884 founded Marfa. The next year Dean succeeded in having the county seat of Presidio County moved from Fort Davis to Marfa, where in 1886 citizens built an expensive courthouse.[7]

Among the first sheepmen moving to the Trans-Pecos as part of the boom were Diedrick Dutchover, George Crosson, and Lawrence

[5]1880 U.S. census statistics cited in M. P. Irving, "The Settlement of the Trans-Pecos or Big Bend Country" (Master's thesis, University of Colorado, 1924), p. 29. For different figures see Samuel Lee Evans, "Texas Agriculture, 1880–1930" (Ph.D. diss., University of Texas, Austin, 1960), p. 267.

[6]Cited in Irving, "Settlement of the Trans-Pecos," p. 35.

[7]*Marfa New Era*, August, 1922; Carlysle Graham Raht, *The Romance of Davis Mountains and Big Bend Country*, p. 256.

Haley. Dutchover first came to the Big Bend with the army. While stationed at Fort Davis, he became aware of the possibilities for livestock production. In the 1870s, after leaving the military, he returned to the Big Bend and despite the presence of Indians established sheep and cattle in Limpia Canyon near Fort Davis. It was an excellent spot, with a plentiful supply of water, good grass, and sufficient timber for fuel. The army post not only offered protection from Indians, but also a market for some of the mutton sheep.

George Crosson, a native of Ireland, was only nine years old when he ran away from home. He went to Canada. Later he came to the United States to work in logging camps in Mississippi, in the gold fields in California, and as a freighter for the Confederacy during the Civil War. He first realized the ranching potential of the Trans-Pecos while on freighting trips to Chihuahua. In 1878 he bought sixty rams in San Antonio and hauled them overland in wagons to Fort Davis. He purchased some 1,800 ewes from Milton Favor and herded them in Musquiz Canyon. Later he settled along Goat Creek in present Brewster County. Here there was running water and plentiful game. But it was also an old Apache haunt, and Crosson had to ward off the Indians, who were accustomed to fattening their ponies there, while he grazed, sheared, lambed, and castrated his sheep. He clung to the land. After Crosson died in 1885, his wife, Lizzie, and her sons continued to develop the ranch. They added cattle to the spread, fenced in the range, and otherwise developed the property.[8]

Crosson's close friend and Big Bend neighbor, Lawrence ("Mick") Haley, another Irish immigrant, settled nearby in 1878. Haley, a brusque and hardy bachelor, had raised sheep in Uvalde County before Crosson convinced him to move to the Trans-Pecos country. With the help of two others, Haley drove 3,000 sheep to Burgess Spring, today's Alpine, and herded in that vicinity until he located his large ranch on Calamity Creek south of Alpine, as a neighbor of his old friend. In 1881 Apaches raided his ranch, murdered one of his herders, and drove off some of his sheep. After he reported the incident to local authorities, Texas Rangers under C. L. Nevill and a detachment of

[8]Holt, "Pioneering Sheepmen," p. 22; Utley, "Range Cattle Industry," p. 427; T. C. Davis, "The +IN Ranch: History and Development of a Pioneer Ranch," *Voice of the Mexican Border* 1 (October, 1933): 76–81.

United States cavalry troops pursued the Indians but did not catch them. Although Indian raids declined, Haley had other problems. The late Barry Scobee, chronicler of many Big Bend tales, noted that Haley occasionally had difficulty keeping his Mexican herders. Once, when dynamite crews of the Southern Pacific Railroad were building through Paisano Pass near Alpine, some of Haley's Mexicans, frightened by the continued blasting, fled for the Rio Grande. They did not stop until a friend, another Haley employee familiar with the construction noise, overtook them to explain the situation. He persuaded them to return to their flocks.[9]

Haley was one of the largest and most successful Trans-Pecos sheepmen. Grazing about 15,000 sheep in the mid-1880s, he also was among the first regional wool growers to fence his range, completing the task in 1886. He put fence posts sixteen feet apart and ran seven to ten strands of barbed wire between them.[10] He profited in the early days when wool prices were good and remained high.

In the 1890s he experienced some of the same price problems other wool growers faced. After passage of the Wilson-Gorman tariff (the "Cleveland tariff," it was bitterly called in the Trans-Pecos), which eliminated import duties on foreign wool and thereby, reasoned wool growers, reduced prices for domestic producers, Haley found his wool supplies selling for only 14 cents per pound. He determined to hold his wool rather than sell it at that price. He built some enormous barns to store the commodity, but the price continued down, reaching a low of 6½ cents. After four years he had more than 300,000 pounds of wool. This he accompanied all the way to Boston, the wool marketing and manufacturing center for the United States. When he offered the wool for sale in the Hub City, buyers told him that the market looked a little weak. "The hell it is!" he is reported to have said, and he proceeded to buy additional wool from the Boston dealers. Later, according to at least one writer, he sold it all at a profit.[11]

[9]Barry Scobee, "Highland Country Once Strictly Cattle, Now Vastly Changed," *Sheep and Goat Raiser* 22 (November, 1941): 11; Roy D. Holt, "Pioneer Cowmen of Brewster County and the Big Bend Area," *Cattleman* 29 (June, 1942): 18–19.

[10]Scobee, "Highland Country," p. 11; Holt, "Pioneering Sheepmen," p. 22.

[11]Holt, "Pioneering Sheepmen," p. 22; Roy D. Holt, "Early-Day Stockmen on the Lower Trans-Pecos," *Cattleman* 24 (November, 1937): 11–14; Madison, *Big Bend Country*, pp. 149–50.

Upon returning to his Big Bend ranch, Haley sold off his sheep and, like many other area ranchers, turned to cattle raising. When he died in 1907, Haley left behind a ranch of 37,000 acres and a personal fortune of one million dollars. The wool storage barns he had erected several years before became the nucleus of the Big Bend's first dude ranch, run by Pete Crawford, who turned one barn into a large dance hall.

Charles Downie, a Scotsman, was another pioneer sheepman in the Trans-Pecos. A man of courage and vision, in 1881 he drove a flock from San Antonio westward to settle briefly along San Francisco Creek in present Brewster County. His only possessions at the time were his sheep and wagon with a team of horses. But he persevered in the arid region, marketing his wool through San Antonio agents and building up his flock. Later he moved into Pecos County to locate a headquarters ranch some fifteen miles northwest of Sanderson. Living in a tent and finding it necessary to haul his drinking water forty-five miles, he nevertheless accumulated more sheep and additional land. To protect his sheep from predators, he erected rock pens, some of which still stand on the old ranch. For a time he penned the sheep every evening. But even then it was necessary to maintain fires all night within the enclosures to keep wolves and mountain lions from destroying the sheep. Since hauling water to his expanding flock became impractical, he instructed his hired hands to drill a well. They did not reach water until the well was three hundred feet deep. After the 1894 "Cleveland tariff," Downie added cattle, horses, and goats to his stock. At one time he counted 80,000 sheep, 20,000 cattle, 2,000 goats, and 500 horses grazing on some 300 sections of well-improved land.[12]

The founder of the town of Marathon, Captain A. E. Sheppard, in 1881 brought sheep to the Trans-Pecos. Having first come to the region with a surveying party working for the Southern Pacific Railroad, he liked the country and traded an interest in a fleet of seven Great Lakes freighters for the Iron Mountain ranch north of Marathon. He purchased his animals in California to ship them to Texas. At one time he had 27,000 head on his ranch. He profited from a large wool clip but switched to cattle after the 1894 tariff wrecked the western sheep business.[13]

[12] Holt, "Pioneering Sheepmen," p. 23; Holt, "Early-Day Stockmen," pp. 11–14.

[13] Utley, "Range Cattle Industry," pp. 429–30; Madison, *Big Bend Country*, p. 148; Holt, "Pioneer Cowmen," p. 19.

In the 1880s John Humphris, known as Don Juan, came from England to Texas. A well-to-do English stock raiser, he sold his farm and put all his money into cattle and horses. These he placed on a ship to carry to America, where he planned to start a ranch. When nearing Corpus Christi on the Texas Gulf coast, the ship sank, and Humphris landed in Texas broke but happy to be alive. He worked on the Richard King ranch in South Texas, taking sheep in payment for his labors. After collecting a sizeable flock, he moved to Maverick County, where he remained for about five years. In 1883 he drove a flock of 6,000 sheep to Presidio County. There he allied with Charley Murphy and Jim Walker to establish one of the largest sheep ranches in the region. In the 1890s Humphris and his partners shifted to cattle.

There were many other pioneer sheepmen in the Big Bend. T. S. Brockenbrow ran sheep in Ranger Canyon. Tom O. and Daniel Murphy raised sheep west of Alpine. Cal Nations and Joe Irving herded in Musquiz Canyon. Sam Salzbury kept sheep southwest of Burgess Springs. Dan Weaver located a ranch in the Glass Mountains. Judge W. W. Bogel owned sheep in Presidio County. Arthur G. Anderson maintained vast holdings in Pecos and Mitchell counties.

As these and other sheepmen entered the Trans-Pecos, they found much of the land empty and unclaimed. Cattlemen could be found at some of the best grazing sites, but plenty of land remained vacant. The pioneer sheepmen settled along creeks and at water holes, claiming all adjoining land. They often herded the sheep themselves at first, but when the flocks increased in size they got help from remittance men, Basques, and American adventurers. Most of them employed significant numbers of Mexican herders, who came north looking for work. Except at lambing and shearing time, when extra help was needed, there was little labor shortage. When a shortage developed, many ranchers traveled by train to Del Rio, crossed the river to Mexico, and hired herders in the bars and saloons of Ciudad Acuña. Farther north ranchers hired Hispanic herders at Fort Stockton or at Pecos. *Pastores* came south from New Mexico ranges to seek employment.

Sheep camps in the Trans-Pecos were not unlike those elsewhere in Texas. They were plain, simple, and unobtrusive. One wool grower remembered that "about all an old herder had was an old bed roll and his cooking outfit. He always kept his personal belongings together in some sort of little bag or box and we usually gave him some sort of tent." Another old sheep rancher reported that "many a time I've gone to [a herder's camp] and sat down on old bleached cow skulls. You often saw those sitting around the camp for chairs." He noted that the skulls "made rather nice stools to sit on, for at least, they were clean and white."[14]

The herders were excellent cooks. They had supplies of beans, flour, meal, and other items, and ranchers usually allowed them to kill a sheep or goat. The herders saved the tallow for lard and always had plenty of meat on hand, though mostly dried. They seldom wasted anything. To prepare beans, the herder put them on the fire at night to let them barely simmer. In the morning he heaped hot ashes and a few coals around the bucket before he started out with the flock. When he returned to camp in the evening, he found the beans "mealy" and tender. He then cut up dried meat in a skillet, added a little onion and a tiny piece of garlic, and cooked the concoction in gravy. Florence Fenley recalls that the beans a herder cooked in his self-sealing bucket were the best beans she ever ate.[15] When supper was over, the herder carefully washed every dish and skillet. Then he washed his dishrags and hung them up on a limb.

The herder was alone much of the time. With a dog or two he guarded his flock of 1,500 to 3,000 sheep twenty-four hours a day, seven days a week, nearly year round. He came to the home ranch only at shearing and lambing time. A rustler brought the herder his rations of food, coffee, tobacco, and other supplies. The rustler also directed the herder as to where and when he should move his flock. Because of the sparse grass in the region, most Trans-Pecos stockmen moved the sheep camps often. When it was time to change ranges, the rustler told the herder where to bring the flock that evening and took the wagon and team to the designated place. Sometimes the camp was near a water hole. But most times the rustler, when he helped to change camp sites, had to haul water in barrels and kegs to the new location.

At sheep camps along the lower Pecos the "honeywagon" would appear. This was an old Mexican hack drawn by a couple of burros and driven by a man. In the back, perhaps hidden by the roll-up canvas sides, perhaps not, sat four Mexican girls, "sure-enough good looking

[14] Florence Fenley, "Sheep Camps and Old-Time Mexican Herders," *Sheep and Goat Raiser* 25 (August, 1945): 8.

[15] Ibid., p. 10.

William L. Black's meat can labels of the 1890s. Courtesy, Southwest Collection, Texas Tech University.

Live Oak Park on William L. Black's ranch, near Fort McKavett. From *Sheep and Goat Raiser* 30 (July, 1950): 5.

Sheep ranch near Devils River. From U.S. Department of Agriculture, *Special Report . . . Condition of the Sheep Industry*, 1892.

Sheepherders' camp. From U.S. Department of Agriculture, *Special Report . . . Condition of the Sheep Industry*, 1892.

Dipping sheep in a tank in southwestern Texas. From U.S. Department of Agriculture, *Special Report . . . Condition of the Sheep Industry*, 1892.

Sheep-shearing scene. From U.S. Department of Agriculture, *Special Report . . . Condition of the Sheep Industry*, 1892.

Freighters hauling wool in San Angelo country in early 1900s. Courtesy, Southwest Collection, Texas Tech University.

The Wool Growers' Central Storage Warehouse, San Angelo, in the days of wool shipments by wagon train. Courtesy, Southwest Collection, Texas Tech University.

Wools arriving for sealed-bid sales at the grower-controlled warehouse in Del Rio. Courtesy, Southwest Collection, Texas Tech University.

Heavily loaded wool wagons in front of Richardson Wool Warehouse, San Angelo. Courtesy, Fort Concho Museum, San Angelo.

Angora goats browsing on Edwards Plateau, 1920s. Courtesy, Southwest Collection, Texas Tech University.

French Rambouillets on Arthur G. Anderson's Hat A Ranch, in Mitchell and Pecos counties, circa 1904. Anderson, one of the first to stock large numbers of the breed in Texas, is on the far right. From *Sheep and Goat Raiser* 26 (December, 1945): 43.

Left: T. A. Kincaid, a major promoter of the Texas sheep and goat industry. From *Sheep and Goat Raiser* 30 (January, 1950): 20. *Right:* Four men largely responsible for founding Texas Sheep and Goat Raisers Association: *top left,* B. M. Halbert; *right,* Virge A. Brown; *bottom left,* R. E. Taylor; *right,* J. B. Murrah. Courtesy, Southwest Collection, Texas Tech University.

Group outside the old Princess Theater in Del Rio the day the Texas Sheep and Goat Raisers Association was organized, in 1915. Courtesy, Southwest Collection, Texas Tech University.

Past presidents of Texas Sheep and Goat Raisers Association, 1915–45. Courtesy, Southwest Collection, Texas Tech University.

girls . . . to a sheepherder," recalled on old-timer who had worked for Arthur Anderson. While they occasionally approached the lonely and isolated camps, more often the honeywagons showed up at the lambing and shearing camps. Asked his business, the man replied that he was a merchant. He offered for sale such items as shirts, hats, tobacco, candy, and "O.K." shoes (a type favored by many herders). The shoes had steel tacks in the soles and a steel strip around the heel, enabling them to stand the heavy wear of ceaseless walking. The "merchant" also pointed to the girls and announced, "*pinche*," a Mexican slang word for prostitution. The girls charged, depending upon several circumstances, between two and six dollars. Herders seldom turned them away. Sometimes at the appearance of the honeywagon ranchers extended an advance to the herder's salary. The pimp and his whores stayed in camp as long as there remained both money and demand for their services.[16]

The number of sheep camps in the region increased as the Trans-Pecos population increased. Concomitantly, ranchers found it necessary to secure title to the land by lease or purchase; the old scheme of herders on free and empty range began to disappear. The state held many of the dry sections as school land, but wool growers experienced little need to contract for those lands until fencing began. Since railroads held most of the sections that contained water, stockmen secured those first, and they could then pasture their sheep and cattle on the adjacent unwatered grassland.[17] Without access to water some of the late-arriving wool growers had little choice but to drift their sheep, following the rains that filled temporary water holes. They often, by design or accident, entered property under lease by cattlemen. Friction followed, but since many cattlemen in the Trans-Pecos also grazed sheep, it was seldom violent.

M. H. Kilgore experienced some difficulty. After agreeing to take charge of a sizable flock for the J. M. Shannon Company, a large sheep outfit with holdings scattered over West Texas and southern New Mexico, Kilgore secured a lease for six sections of land near the Chisos Mountains south of Marfa. He located his first camp at Rosillo Spring on the west side of the mountains. About the third day, three cowboys

[16] Philip Thompson, interview with Elmer Kelton, Philip Thompson, Papers, Southwest Collection, Texas Tech University, Lubbock.

[17] Utley, "Range Cattle Industry," p. 433.

rode in demanding to know what he was doing in their territory with sheep. "I saw," he reported later, "that I was outnumbered so I resorted to strategem [*sic*]." Since he had been reading Jules Verne's *From Earth to the Moon*, a mid-century science-fiction tale about a space flight to the moon, Kilgore pretended that he was the French author and had taken the trip. He told the cowboys that he and his partner had built a machine to make the trip. After they got the contraption loaded with ballast, water, supplies, and their dog, they took off about dusk. After reaching a height of 5,000 feet, about midnight, he told the cowhands, they began losing altitude. They began throwing out ballast. That did not help. They threw out the supplies. Finally, they threw out the dog. They then traveled without further difficulty. At daylight, they looked out and discovered the dog traveling beside them in the air.

When Kilgore got to that point in the yarn, one of the cowboys said to the others: "This damn fool is crazy. Let's go."

"I didn't mind seeing them go just then," remembered Kilgore, "but later on I got to be friends with those fellows and we talked about the first visit of theirs a lot and laughed over it."[18]

Philip Thompson, a herder for Arthur Anderson, had a similar experience. In the late 1880s Thompson and two others were taking some 6,000 sheep from Anderson pastures in Mitchell County to the stockman's vast Hat-A Ranch near Fort Stockton. After wading the Pecos at Horsehead Crossing, they put the sheep on a greasewood flat. It was school land leased to a cattleman. The next morning they had the sheep spread out along a mile front, pointing them in the direction of Fort Stockton, about twenty-five miles to the southwest. Five cowboys, each with a Winchester rifle on his saddle rode in at a lope. One of them demanded, "Whose sheep are these?"

Thompson said, "They belong to Arthur Anderson."

"Well get them out of here. This is my ranch. I don't want them here."

The herder, who was several years younger than the spokesman for the cowboys, replied; "Well, you just wait a minute, young man. These sheep are spread out, they're grazing, they're going on and on."

"Well, get them on out right now. We'll help you."

[18] Florence Fenley, *Old Timers, Their Own Stories*, p. 22.

"We don't need any help," answered Thompson. "I don't want your help. If you bother these sheep, they'll all run together, and they won't move 'til late this afternoon. I know what I'm talking about."

"That's what you say. Let's help him boys." With that the cowmen fired their Winchesters in the air. Predictably, the sheep bunched up. Then the riders galloped their horses into the sheep dashing one way and then another trying to stampede the timid creatures. Their efforts only drew the sheep into a knot. After what seemed an hour to Thompson the intruders rode off, leaving the sheep to the herders.[19]

In the late 1880s drifters—herders without a lease to the land their sheep grazed—proved a problem for settled sheepmen and cattlemen. Coming out of New Mexico in the early spring with flocks of 1,500 to 2,000 animals, drifters followed the Pecos Valley to about Horsehead Crossing, where they cut south toward the Rio Grande. They followed the rivers and creeks in the hot summer before turning back north to reach the home ranch for the winter. They grazed their sheep leisurely, searching for the best grass and reliable water supplies. When they located a favorable range, they sometimes camped out for weeks. If cowboys or rustlers rode into camp to hurry the intruder on his way, the drifter, who in preparation for such a contingency had removed one of his wagon's good wheels, hidden it, and put a broken one in plain sight, said he would move on as soon as he had repaired the wagon. When his animals had depleted the range, he retrieved the good wheel, placed it on the wagon, stashed the broken one inside to use again, and hastened on his way to another verdant range.

In the Trans-Pecos the first sheepmen marketed their wool through agents representing eastern textile firms. After railroads pushed through the region, they hauled it to such local stations on the Southern Pacific as Valentine, Marfa, Alpine, Marathon, and Sanderson, from where it went by rail to Boston, New York, Philadelphia, or other East-Coast cities. Most of the lamb crop they kept or sold locally. Older sheep, used as mutton animals, they shipped to stockyards at San Angelo, Del Rio, El Paso, Fort Worth, or San Antonio.

As the Trans-Pecos filled with sheep and cattle operations, ranchers began fencing their land. Haley and Crosson were among the

[19]Thompson, interview with Kelton, Philip Thompson Papers.

first to fence, but others were not long in following. Drift fences on cattle ranches went up first. Ranchers later completely enclosed ranges. By about 1892 or 1893 the Trans-Pecos stood crisscrossed with barbed-wire fences. By this time, too, farmers had entered the region, planting wheat, corn, and cotton. The northern counties accounted for most of the farmers. Areas near Balmorhea, Toyah, and Pecos in Reeves County attracted many wheat and cotton growers. The upper Rio Grande Valley also became popular for raising crops.

In the 1890s two events cut heavily into sheep and wool production. A severe drought in 1892 and 1893 emptied the water holes and creeks in the sheep-producing sections. Wells dried up. Grass on the rangeland shriveled, browned, and turned under itself. Some livestock died. Some wool growers sold their animals rather than see them starve. Some continued to handle sheep, but succeeded only by hauling water at an enormous cost, over a great distance, and at considerable inconvenience. It was not a pleasant experience. Then, as described above, in 1894 Congress passed the Wilson-Gorman Tariff, which reduced wool prices to as low as 6½ cents per pound, too low for wool growers in the Trans-Pecos to make a profit. Many of the wool growers sold off their sheep and turned to cattle raising.

Not all sheepmen gave up. Some stuck it out, pressuring their representatives in Washington to repeal the 1894 tariff or to write a new one. In 1897 Congress restored wool protection to the tariff schedule. Meanwhile, low prices for land and sheep attracted a few adventurers and promoters who hoped to take advantage of hard-pressed livestockmen. Indeed, depressed ranching conditions in the 1890s brought to the Trans-Pecos one of its most colorful characters—Dominicker Hart.

Dominicker Hart was born in Ireland. Coming to Texas in the early 1880s, this man of ready wit and great business acumen accepted a job herding sheep near Carrizo Springs, county seat of Dimmit County. He took sheep on shares, saved his money, built a small flock, and added more by purchase. He was a promoter. One year he sold his sheep just as the price began to slump. The next year he bought them back for less than half the price he had received. With the profits he added more sheep. He was a gambler. One time in a poker game he bet his sheep against another flock. When he won, he launched into the sheep business on a grand scale. For a few years he had several flocks attended by herders near Del Rio. He kept such close tabs on his operation that he himself acted as rustler, oversaw the ranges, and went to town to purchase supplies.

In 1895 when the region's sheep industry was at low ebb, Hart moved into the Trans-Pecos. He established a headquarters ranch five miles north of Pumpville, along the Southern Pacific Railroad in far west Val Verde County. He operated on an enormous scale, his range eventually covering practically all the territory between the Pecos and the Rio Grande and extending westward to Sanderson Canyon in Terrell County. He controlled more than 1,000 square miles. Over this vast territory he located sheep camps at some of the springs and a few scattered wells. He also raised cattle, horses, and goats.

Sheep raising was the most important phase of Hart's empire. At one time he owned more than 100,000 head, selling in some years as many as 30,000 animals. He sheared up to 90,000 sheep a year. He employed nearly sixty men for the sheep work, doubling that number at lambing time. Shortly after the turn of the century, he accompanied a shipment of his wool to Boston. He planned to corner the American wool market. He had twenty-seven railroad boxcars loaded with his own wool and held certificates for additional lots of wool. In Boston he purchased more wool, but to no avail. His scheme failed. The New England Yankees were too much for him, he admitted, and he lost a quarter of a million dollars. Back in Texas, he continued his extensive ranching, buying or leasing all the land he could get. But after the Boston imbroglio, he concentrated on horses. When he sold out in 1916 to retire in Fort Worth, he still held 80,000 head of sheep. For these he received an average of six dollars apiece. He lived comfortably in retirement.[20]

Despite the individual successes of men like Dominicker Hart, the drought and the "Cleveland tariff" of the 1890s temporarily weakened wool growing in the Trans-Pecos, and ranchers shifted to cattle production. From a high of an estimated 174,000 in 1893 the number of sheep slipped to less than 145,000 two years later, far below the number of cattle. By 1900, however, there were 381,000 sheep in the re-

[20] Newspaper clipping, Morgue Files, Texas Sheep and Goat Raisers Association, Records, Southwest Collection, Texas Tech University; Roy D. Holt, "Dominicker Hart Made Sheep Pay West of the Pecos," *Sheep and Goat Raiser* 30 (June, 1950): 49; Hiram Phillips, interview with Paul H. Carlson, October 1, 1981.

gion, with some sheep in each county, varying in number from 21 in Jeff Davis County to 142,764 in Pecos County. As ranchers continued to graze sheep and goats in the area, the number gradually increased again. But not until the 1930s did the Trans-Pecos sheep industry seriously challenge the predominance of cattle.

The early sheep and goat industry in the Trans-Pecos was similar to that in other parts of West Texas. It was an extension, in many ways, of sheep and goat raising on the Edwards Plateau. It boomed in the 1880s at the same time as it enjoyed rapid growth on the Edwards Plateau and in the Rio Grande Plain. It differed in that landholding in the Trans-Pecos was much larger on average than in other parts of the state. It differed also in that many more ranchers in the Trans-Pecos, following the lead of Milton Favor, stocked both sheep and cattle on their ranges than ranchers elsewhere. As a result, the sheep wars in the Trans-Pecos were not nearly as serious as elsewhere.

11
The Sheep Wars

THE Texas sheep wars extended from the arrival of the first Spanish *chaurros* and their herders to a period at the end of the nineteenth century when the sheep and goat industry underwent a transformation from an adventure to a settled occupation. The wars were characterized by Indian attacks on the sheep and the men who guarded them, by Rio Grande border raiders who struck across the river from bases in Mexico, by confrontations between Mexican and American teamsters hauling wool, and by fighting between cattlemen and sheepmen on the West Texas frontier. The wars included fence-cutting, arsons, theft, intimidation and beatings, legal action, and murder. While there were violence and bloodshed, the Texas sheep wars produced no such lasting armed conflicts as occurred in Arizona, New Mexico, Wyoming, and Montana.

Violence over Texas sheep evolved through several phases. During the Spanish period, Indian raids on mission livestock and nearby civilian flocks dominated the troubles. From the early American period through the Mexican War, Indians, fighting for their land and for food, attacked isolated farms and settlements on old empresario grants. After the Mexican War ended in 1848, the sheep conflicts developed along two fronts: Indian and outlaw border raiders from Mexico crossed into Texas along the Rio Grande, and Indians struck at isolated points on the western frontier. These raids continued until the mid-1880s. The final phase of the Texas sheep wars was the antagonism between sheepmen and cattlemen in the open-range country of West Texas. With the population increasing and free land becoming scarce, the final phase represented the most troubling period. In the conflicts sheepmen, although sometimes the assailants, were the chief victims.

From almost the first appearance of sheep in Texas, Indians raided sheep camps. They chased and killed the animals for food and fun. Resenting the appearance of white men in their midst, Indians attacked Spanish mission stations to drive off cattle, horses, sheep, and goats. They struck at herders, burned fields, and otherwise prevented an easy existence for the Spanish intruders.

For the Spanish sheepmen in the eighteenth century the wide-ranging Comanches and Lipan Apaches proved especially troublesome. In one attack Apaches killed José Gonzales and two of his Indian neophytes before butchering fifty sheep and scattering the rest. In another, Indians on foot drove a flock of 1,500 sheep over a steep precipice, killing or injuring nearly all of them. Unhealthy conditions and Apache attacks, including the death of Father Gonzobal, caused the removal of the San Xavier mission from near present-day Rockdale. Indian attacks prevented the establishment of missions in northeast Texas above a line running between San Antonio and present Natchitoches, Louisiana. The Spanish abandoned a mission and presidio near modern Menard with more than three hundred colonists settled nearby. Athanase de Mézières reported that Indians in the San Antonio district ran off sheep and drove herders to cover at almost regular intervals.[1] In East Texas Frenchmen and Indians stole sheep and cattle and otherwise molested the livestock operations. Indian hostility was fearsome enough to play an important factor in the decision after 1767 to abandon most mission fields in Texas.

In the early nineteenth century, as Americans entered Texas, Indians continued to resist encroachment on their territory. They pounced upon outlying settlements, stampeding cattle and running off sheep. They invaded sheep pens. Settlers as far south as the John McMullen and James McGloin grant at San Patricio endured repeated Comanche raids on their sheep.[2] Elsewhere Tonkawas, Karankawas, and Wacos stole small numbers of sheep. Indians destroyed or scattered 500 sheep of W. G. Slayton, east of present Waco, as he drove the animals toward settlements within the Peters' Colony grant. Joe B. Martin watched Indians butcher ten sheep he owned near Gonzales. West of Corpus Christi settlers repeatedly endured Indian assaults on

[1]Herbert Eugene Bolton, ed., *Athanase de Mézières and the Louisiana-Texas Frontier, 1768–1780*, II, 118–24, 240–42.

[2]C. A. Gulick, Katherine Elliott, and Harriet Smither, eds. *The Papers of Mirabeau Bonaparte Lamar*, III, 515, and V. 379.

their livestock, including sheep. The pioneers fought back, however, and the losses were small enough that they never really slowed American settlement.[3]

During and after the Texas fight for independence many Mexican herders, driven by ethnic prejudice against them and a general lawless state, fled isolated points north of the Nueces River. They led their flocks southward to below the Nueces, staying near the Golf coast, toward the Rio Grande. Some of them disposed of their animals or trailed them to points beyond the Rio Grande. This southward displacement of sheep raising continued during the War with Mexico, effectively emptying a once occupied range. In the wake of retreating Mexican herders, Germans and Americans carried in sheep, thus maintaining and even increasing the numbers north and east of the Nueces.

After the succesful conclusion of the War with Mexico, which secured the Rio Grande as the southern boundary for Texas and the United States, sheep wars of two types developed. On the western frontier of settlement Indians fought for rights to their land, and along the Mexican border outlaws, malcontents, Mexican patriots, and Indians raided sheep operations for plunder, food, and revenge. The western-frontier Indian raids are fairly well documented. Many Texans recorded their experiences with sheep and Indians. George Wilkins Kendall, far out on the frontier near present Boerne, found it necessary to maintain a system of armed sentries every night to ward off Comanche pressure. For each flock of 800 sheep he provided a herder with a double-barreled shot gun, a Bowie knife, and a Colt six-shooter. The scheme worked. Other frontier sheepmen faced similar inconveniences. A neighbor living northeast of Kendall hired guards and asked his brother in Kentucky to bring several dogs with him when he came to the Texas Hill Country. During the Civil War, which drew away many adults, Indians became ever bolder in their attacks. Sheepmen sometimes found their herders dead, slaughtered under a veritable avalanche of arrows, and their sheep run off. Some sheep producers lost their entire flocks to Indians. Kendall noted that a period of terrorism

[3]Edward N. Wentworth, "The Golden Fleece in Texas," *Sheep and Goat Raiser* 24 (December, 1943): 16, 20; Gulick, Elliott, and Smither, *Papers of Lamar*, V, 379; T. R. Havens, "Texas Sheep Boom," *West Texas Historical Association Year Book* 28 (1952): 4.

swept across the Texas frontier, forcing at times his own family to barricade itself in the ranch house.[4]

Henry Shane, like Kendall a Texas pioneer sheepman, experienced repeated trouble with Indians. In 1850 Shane went to work for a Major Baredon, quartermaster at Fort Clark near Brackettville in present Kinney County. Taking charge of Baredon's sheep in the vicinity, Shane supervised the work of several Mexican herders. One day Indians surprised, captured, and severely whipped him with a pair of hobbles. They placed him on a mule, and the party traveled rapidly all day and part of the night. When it stopped, the Indians whipped him again. But here a trailing party attacked the Indians, and Shane managed to escape. He returned to Fort Clark to continue in the sheep business. In a second attack Indians killed two of his herders. He carried on, however, undauntedly facing additional Indian dangers. As late as 1870 Shane and his brother suffered occasional Apache raids, one time barely escaping alive.[5]

Along the Rio Grande the problems were different, but the results were the same. After the war between Mexico and the United States, Mexico was in a constant state of turmoil. Along the river as well as throughout the vast borderlands, even though there were many instances in which Mexicans and Americans formed warm personal friendships, there was in general a deep hostility between peoples of the two countries. Perhaps the bitter feeling was a natural outgrowth of the Texas revolution and the subsequent war, which saw Mexico lose nearly half its national territory. The mass of the Mexican people regarded Americans with dislike and suspicion, worrying that the Americans might take more of their country. Americans believed that Mexicans were treacherous, cruel, and undependable.[6] From Mexico outlaws, cattle and horse thieves, Indians, and others crossed into Texas, killing, stealing, and running off livestock before fleeing back to their sanctuaries below the river.

One of the most famous of these raiders was Juan Neponuceno

[4] Wentworth, "The Golden Fleece in Texas," 20. See also Harry J. Brown, ed., *Letters from a Texas Sheep Ranch*, pp. 1–20.

[5] Roy D. Holt, "Pioneering Sheepmen," *Sheep and Goat Raiser* 26 (December, 1945): 23.

[6] See, for example, Clarence C. Clendenen, *Blood on the Border: The United States Army and the Mexican Irregulars*, pp. 17–18.

Cortina, a prominent Mexican-American who was looked upon by Mexicans as a kind of modern-day Robin Hood. Headstrong and willful, this flamboyant character with a red beard and gray eyes was from a wealthy family, but he preferred the company of *vaqueros* and *pelados* ("the poor"). He built up a considerable following among Mexicans of the lower Rio Grande borderland, many of whom had genuine grievances against the Americans. In the late 1850s and afterward he led, or legend maintains that he led, vicious forays throughout the lower border, striking as far north as Corpus Christi. He and his men killed herders, burned houses and corrals, and ran off livestock. Quoting from a government report on the attacks, John Ashton writes that from October, 1859, through January, 1860, Cortina and his gangs at Mifflin Kenedy's ranch in Hidalgo County ran off or destroyed:

1,576 head of horned cattle, at $7	$11,032
400 head of beeves, at $20	8,000
2,400 head of improved goats, at $1	2,400
18 head of horses, at $40	720
1 mule	60
improvements, frame house, corrals, etc.	1,200
	$23,412[7]

Ashton found that several other sheepmen around the lower Rio Grande and Gulf coast complained of the destruction of livestock. Most of them attributed the work to Cortina. Ashton notes that Thaddus M. Rhodes of Brownsville announced the loss of, among other livestock, 90 head of sheep and goats, valued at $1.50 each. W. Neale and Nestor Maxan, who jointly owned a ranch known as the Baston a few miles from Brownsville, lost on October 30, 1859, among other property, 200 graded goats and sheep, which they valued at $4.50 per head. William Johnson, Cameron County, included among his lost possessions 70 sheep at $2.00 per head and 90 goats valued at $1.50 each. Antonio Dodies lost 100 head of sheep valued at $100.00. Peter Champion, Cameron County, reported that 50 sheep and 100 goats were carried off during October and November, 1859. Nicolas Champion lost 200 goats valued at $1.50 per head. Two men, named Doddridge and Jacobs, of Starr County reported losses that included $30,000 worth of

[7] Lyman L. Woodman, *Cortina, Rogue of the Rio Grande*, pp. 3–25; John Ashton, "The Start of Sheep Breeding in Texas," *Sheep and Goat Raiser* 25 (December, 1944): 36–37.

hides, wool, and specie. Afterward, in search of greater security, they emigrated to Mier, Mexico.[8]

The border incursions prompted American military action. Texas Rangers and sometimes troops of the United States Army struck without success at what they believed to be Cortina's bases in Mexico. During the Civil War, when American military defense was unavailable for ranchers in the lower Rio Grande, the persistent Juan Cortina stepped up his international murder and theft. As a result Anglo sheepmen abandoned their ranches in South Texas. After the war Cortina continued to raid periodically until the mid-1870s. His raids caused considerable animosity.

This kind of animosity, generalized toward Mexicans, played a role in the 1857 Cart War. A lively trade between San Antonio and the Gulf coast at Corpus Christi and Indianola included the overland use of the large two-wheeled Mexican carts and the traditional Anglo four-wheeled wagons. An intense rivalry sprang up between the Mexican carters and the American wagoners. The teamsters hauled all kinds of goods, including wool from collection centers at such places as Eagle Pass and Laredo, to a small but growing San Antonio market and thence to the coast. The Mexican carters transported their cargo more cheaply than the wagoners and therefore did most of the hauling. To weaken Mexican competition, a resolution passed in Uvalde County prohibited Mexicans from traveling through its jurisdiction. Undaunted, the carters continued their trade. Fighting followed. The war was characterized by a series of attacks by mounted Texans, disguised and armed, against the cart trains along the busy public road between San Antonio and the coast. In Karnes County in August the Texans killed at least one carter, José Antonio Delgado. Before Texas Rangers quelled the conflict in December, several other Mexicans had died and others had been wounded.[9]

During and after the Cart War Indian and outlaw bands from Mexico struck ranches in Texas. Rustlers operating from bases in Mex-

[8] Ashton, "Start of Sheep Breeding," p. 37.

[9] Ernest Charles Shearer, "Border Diplomatic Relations between the United States and Mexico, 1848–1860" (Ph.D. diss., University of Texas, Austin, 1939), p. 224–27; Alice Freeman Fluth, "Indianola, Early Gateway to Texas" (Master's thesis, St. Mary's University, 1939), p. 39; Walter P. Webb and H. B. Carroll, eds., *The Handbook of Texas*, I, 302.

ico stole livestock north of the Rio Grande and herded the animals southward across the river to sell to Mexican *hidalgos* ("noblemen"). They took cattle, sheep, goats, and horses. Indians, especially the uprooted Kickapoos and the fierce Lipan and Mescalero Apaches, waded the river to run off livestock for food before fleeing to their village south of the Rio Grande. Their raiding activity occurred frequently during the Civil War years and continued at a lesser but troublesome pace afterward.

To combat the raids, the United States Army stationed large numbers of troops at posts along the Rio Grande from Brownsville to Laredo. The troops enjoyed only mixed success. After a dramatic 1873 thrust at their base in Mexico led by Ranald Mackenzie, the Kickapoos no longer penetrated Texas, and the border region below Laredo was largely secured. But the upper Rio Grande border region from Laredo through present Big Bend National Park to old Fort Leaton proved another matter. Although adjacent regions on both sides of the river were almost uninhabited, the region served as a natural haunt and even highway for bands of Lipans and Mescaleros who slipped across the Rio Grande at any one of a dozen crossing places and moved through the chaparral until they reached a place far from the river, where sheep, goats, cattle, or horses grazed. Any luckless herder or traveler who got in the way they killed. With hard riding, the Indians with their captured livestock returned to the river and vanished, often before news of their foray became known.

According to the lucid report of an army officer stationed in the area, on Sunday, April 14, 1878, a band from Coahuila composed of about forty Lipans, Mescaleros, Mexican desperadoes, and at least one American outlaw waded the Rio Grande near the base of Apache Hill about forty-five miles northwest of Laredo. Within minutes of their arrival in Webb County the vicious bandits had killed two *vaqueros* and hurried southeastward along the main road that meandered down the Rio Grande toward Laredo. At dusk they killed Gorgea Garcia, a Mexican herder, within hearing distance of his wife and friends on his ranch a few miles from Apache Hill. After stealing Garcia's goatskin leggings, horse, saddle, and other *chivarras* ("chaps"), as well as a nearby drove of gentle horses, and stampeding his sheep, they resumed their course down the river. Two days later, when within fourteen miles of Fort McIntosh, near Laredo, the raiders turned sharply northeastward toward the Nueces River. After stealing from Dr. Henry Spohn's ranch "a sufficiency of horses to mount all their party," they divided into smaller predatory groups. Their organization seemed perfect. They struck simultaneously at almost all the ranches in the vicinity, stealing the best horses and devastating the territory in a broad sweep. They killed *pastores*, butchered sheep, drove off cattle, and burned ranch homes. After the rapid strikes, they reunited near the Ranchos de los Machos in Webb County, where they wounded Tomás Solís, a cowboy, and then rode on toward old Fort Ewell on the Nueces River in La Salle County.[10]

The terrifying marauders had selected a propitious time for their raid. The verdure of spring had clothed the earth with abundant grass. Shrubs, bushes, and trees enjoyed full green foliage, and the ponds overflowed with water. Because ranchers of the vicinity were busy shearing their flocks or working cattle roundups, few people moved about to note the outlaws' progress. During the day large trees with heavy foliage screened them from observation, and the full moon lighted their way by night. With grass for their horses abundant and water available everywhere, the bandits traveled in nearly a straight line for sixty miles to Fort Ewell.

From the abandoned military post, the raiding party descended the Nueces River for fifteen miles to a point at which it swings northward in McMullen County and arrived there early on the seventeenth. This place had long been a favorite rendezvous for raiders from Mexico. Here, the bloodthirsty bandits in an exultant and unrepressed demonstration of joy celebrated their recent success. Perhaps drunk from too much liquor, they "held high carnival." They executed John Steel, massacred Mrs. William H. Steel's two sons, eight and twelve years old, by mangling and mutilating them with knives, and killed two *vaqueros*, Martin Martínez and Florenzo Leo. Here also they fatally wounded Venturo Rodríquez, a *pastor*, with a rifle bullet and eight arrows and stripped two other *pastores* naked and compelled them to run foot races before murdering them.

[10]Captain John O. Elmore, Twenty-fourth Infantry, at Laredo, to Acting Assistant Adjutant-General, District of the Rio Grande, Fort Brown, May 23, 1878, in U.S. Department of War, "Affairs on the Rio Grande and Texas, 1875–1881," 4584 Adjutant General's Office (AGO) 1878, Letters Received 1805–1889, Record Group 94, National Archives.

After gorging themselves on stolen beef, the marauders galloped southeastward about twelve miles, then changed their direction to almost directly south. At noon on the eighteenth, they struck Rancho Solidad in Duval County, about thirty miles from San Diego, where they killed Guadalupe Basan, a cowboy, before gathering all the horses in the neighborhood. Soon afterwards, on their way toward the Rio Grande, they killed another herder and his wife and butchered nearly three hundred of his stock. The next morning they mortally wounded Margarito Rodríquez, a *pastor*, and stampeded his flock.

When citizens of the vicinity, led by Frank Garvis, rode in pursuit, a sharp skirmish occurred. But the bandits compelled the Garvis vigilantes to retire to a safe position. The brigands then fled toward Mexico, killing enroute several herders who were shearing their flock. They reached the Rio Grande at a point twenty-five miles below Laredo late at night on the nineteenth and escaped with their plunder to the safety of foreign soil.

The raiders, who covered over 265 miles, had been in the United States for six days. During that interval, they had been within the vicinity of government troops at Fort McIntosh, once within 14 miles and again within 25 miles, had passed within 30 miles of federal cavalry troops at San Diego, and had at no time during the week been more than 60 miles from a United States military post. Nevertheless, they had escaped the notice of the army. Well-armed with Spencer and Remington rifles, the brigands had killed at least eighteen people and stolen nearly two hundred horses, rifles, money, clothes, camp equipage, saddles and blankets, and other items. They had burned some buildings and homes, damaged two wells, and killed, stolen, or stampeded more than 2,500 head of sheep and cattle.[11]

Such sanguinary border incursions prompted American military action. On several occasions United States Army troops crossed into Mexico to punish Indians and whites. Sometimes the Mexican Army intercepted the American regulars, but each time the Mexicans turned aside. When another military encounter occurred in Mexico after the bloody 1878 attack in Texas, Porfirio Díaz, the new president of Mexico, who could not afford to have American troops embarrass his army, moved to halt the depredations by controlling the northern frontier.

[11]Ibid.

There was another border difficulty. It involved Mexican herders who drove their sheep northward into the United States in the spring and back to Mexico in the fall. Along the many miles of unguarded river from below Laredo upriver to about present Langtry, Mexican sheepmen after the Mexican War enjoyed unobstructed entry into Texas. Where deep canyons did not block their path, they also waded the Rio Grande into the Big Bend region. Under pressure from Texas cattlemen, the state legislature acted to stop the practice in 1874 by passing an appropriate law. The legislation provided for court-ordered seizure of sheep and goats brought into Texas for the purpose of grazing when the owner was not a citizen of Texas or of some state of the United States.[12] While it stopped the entry of Mexican flocks into populated areas about Laredo, the law had only limited effect farther west in the practically uninhabited Big Bend.

Of far greater concern for wool growers was the antagonism of Texas cattlemen. There is something elusive and contradictory about the sheepmen-cattlemen warfare, for cowmen often ran sheep themselves. On the fabulous King Ranch of South Texas, for example, known for its vigorous Santa Gertrudis breed, Richard King raised nearly 30,000 sheep with his cattle in the 1860s. His cowboys rustled sheep. Later, after the turn of the century, the irrepressible Dominicker Hart, one of America's largest wool producers, grazed thousands of cattle with his huge flocks of sheep. His herders handled cattle. Indeed, most of the Texas sheepmen had cattle. Moreover, such far-sighted promoters as Charles Schreiner, the wool warehouseman and banker at Kerrville, through his unique banking policies encouraged cattlemen to herd sheep. He required cowmen to use part of the money obtained in any loan to purchase sheep.

Adding to the elusiveness of the subject is the fact that wars between sheepmen and cowmen were rare. They occurred; a classic example is the fence-cutting war in 1883. But contemporary nineteenth-century popular culture, historians, and others have exaggerated the fighting. Finally, the mutual antagonism that did exist in Texas was localized in time and place. It did not occur in East Texas or before the second half of the nineteenth century. Nor has it occurred in recent years. Modern Texas livestock raisers know that the sheep and wool

[12]T. R. Havens, "Livestock and Texas Law," *West Texas Historical Association Year Book* 36 (1960): 27–28.

market is more stable than the beef market and accordingly often, but not always, raise both sheep and cattle. And raisers on the Edwards Plateau produce goats and mohair with their sheep and cattle.

Still, cattlemen and sheepmen argued over grazing rights, water supplies, fences, and other matters, particularly between the early 1870s and the late 1890s, when both industries boomed. Sheep and cattle required different amounts of water, different types of food, and different manners of herding. The longhorn was a wild creature, requiring long grass and plenty of water. His very nature required men on horses who could cope with his feral constitution. Sheep, on the other hand, were thoroughly domesticated and subsisted on short grass and weeds. Except in hot weather or when being trailed, they needed little water, but got by on the morning dew. They needed a constant attendant, preferably one who went on foot.

Contrasts in life-styles, background, and equipment also produced hostility and bred disagreement. The Civil War with its regional hatreds accentuated the normal conflict of interests. Cowboys—those not Texans—were immigrants from the southern states who had little experience with sheep. Few herders were Texans or from the South. They were foreign immigrants or Yankees, experienced with wool production, who drifted into Texas. But even before the Civil War many cowmen in Texas had come to the conclusion that herding sheep was no job for a "white" man. They associated it with Mexican *pastores*. Cowboys, moreover, were young (average age, twenty-four) and remained cowboys an average of only seven years.[13] Some were rowdy adventurers or fugitives from the law. Herders were older and on the average remained in the occupation longer. The *pastores* among them were little better off than serfs. The herder went alone and on foot with a dog or two, lived in a tent or in the open, seldom carried a weapon, and dressed miserably. Few were outlaws. The cowboy usually worked with others, dreamed of wealth, and craved excitement. He had an exalted, heroic image of himself as a hard-riding, fast-shooting hombre who broke wild horses, shot it out with Indians, and lassoed bears for fun. He was a practical man of action, who fried his brains under a hot Texas sun for most of the year but often spent the winter months in town. The herder was a philosopher, a "snoozer" as the cowboys called him, who took a nap when his sheep stopped their grazing during the

[13] William H. Forbis, *The Cowboys*, p. 17.

midday heat but who spent the entire year alone on the range with his animals.

There were other differences. Two well-known paintings by famous western artists, if oversimplifying the nature of each, summarize the contrasts. In the Amon Carter Museum in Fort Worth is a canvas of some cowhands by Charles Russell. In the picture a group of young cowboys have snared an enraged grizzly bear with lariats, an event that actually happened. The painting stresses the dramatic instant, with cowboys on lurching horses, guns on hips, neatly dressed in Levis and broad-brimmed hats. The young men are engaged in dangerous, action-packed recreation. The other painting, by N. C. Wyeth, is of a western herder. An old man, rugged and tattered, sits alone under the stars with his sleeping flock, a watchful dog, a dying fire. In one roughened hand is a well-worn pipe, in the other an open book. But the man, staring into the fire, not into the book, seems to be thinking.[14]

The struggle between sheepmen and cattlemen, then, was a conflict of individual and institutional interests. It permitted little compromise or cooperation. Certain myths deepened the hostility. Cattlemen claimed that a thousand sheep would do more damage to the range than ten thousand cattle because they ate the grass close to the ground and because, being closely herded, they trampled it out. The cattlemen also claimed that because sheep gave off a peculiar odor neither cattle nor horses would graze where sheep had ranged. Several factors deepened the misunderstanding. The factions avoided one another. At Fort Stockton, for example, in the 1880s sheepmen traded at one store; cattlemen, at another. They protected members of their own group. Judge H. H. Butz of Pecos County once saw a cowman shoot and kill a sheepman. At the subsequent trial the jury, all cattlemen, acquitted the defendant on the grounds of self-defense. Women proved just as belligerent as the men. Barry Hubbs, son of an early cattleman in Tom Green County, for instance, remembered that once when herders pushed their sheep across the North Concho River onto land his family claimed as theirs, his mother with a shotgun in hand chased the sheepmen off the family's range.[15]

[14] Described in Winifred Kupper, *The Golden Hoof: The Story of the Sheep of the Southwest*, p. 78.

[15] Barry Scobee, "Highland Country Once Strictly Cattle, Now Vastly Changed," *Sheep and Goat Raiser* 22 (November, 1941): 11; Barry Hubbs, interview with Richie Cravens, July 8, 1977, Southwest Collection, Texas Tech University, Lubbock.

Throughout most of West Texas, where the range was open and free, cattlemen arrived first with their herds. They insisted, as a result, that by right of preemption the land belonged to them. Intruders were interlopers, had no rights, and should move on. When sheepmen successfully persisted, trouble started. Sometimes the trouble was a warning like that given to M. H. Kilgore in the Big Bend; sometimes it was an inconvenience like that imposed on Philip Thompson near the Pecos River. Sometimes it was more serious. On one occasion C. F. Adams of Sonora, a former employee of William L. Black, drove a herd of fat steers to a fresh camp on the Pecos. Along the way his outfit passed a sheep camp. When an altercation developed, Adams shot the German herder, wounding him in the leg. Authorities arrested Adams and held him in jail at Del Rio, but later they released him. He never stood trial.[16]

To resist the encroaching flockmaster, cattlemen sometimes formed local protective associations. In early Tom Green County members of such a group established a "deadline" at the North Concho River. They kept their herds north of the waterway and notified sheepmen to keep their flocks south of the stream. When three herders, pushing northward with some 9,000 head, neared the river, the cowmen sent a delegation to remind the sheepmen to stay below the boundary. The sheepmen persisted, however, and crossed the deadline. The local association thereupon called out its members. To avoid being recognized, they covered their faces, using old cloth sacks with eyeholes, and thus gave rise to the term by which sheepmen knew them and most other attackers: gunnysackers. They whopped down on the sheep camp and within minutes killed the owner of the animals and one of the herders. The third herder fled back to the river. The gunnysackers then turned on the helpless sheep, destroying 2,000 head before leaving.[17]

At other times gunnysackers wielding clubs, often taken from the wheels of a sheep wagon, rode through bleating flocks, smashing skulls until they grew arm-weary. At other times they picked off sheep one by one with rifles. Sometimes they used dynamite charges to kill or maim dozens of sheep. More often the gunnysacker set fire to the grass. Since frightened animals instinctively huddled together, this

[16] T. R. Havens, "Sheepmen-Cattlemen Antagonism on The Texas Frontier," *West Texas Historical Association Year Book* 18 (1942): 13.

[17] Hubbs, interview with Cravens, July 8, 1977, Southwest Collection, Texas Tech.

method proved diabolically efficient. One or two flaming sheep could ignite a whole flock of closely herded animals. In another tactic gunny-sackers stampeded a herd of cattle through a flock, trampling and scattering the sheep in every direction.[18]

George McCall of Llano County in 1883 drove a flock of sheep from his territory to Taylor County. As he attempted to water his sheep in Coleman County at a well belonging to Bart Stevens, a cattleman, Stevens ordered him off. Since his sheep had been without water for nearly three days, McCall refused to move the thirsty animals without water. He pointed out that no cattle were in the vicinity and that no harm would come from his watering the sheep. Stevens thereupon rode away, to return a short while later with two of his cowboys. The three men then severely beat up McCall and killed many of his sheep.[19]

The advent of barbed-wire fencing produced most of the "wars" between cattlemen and sheepmen. Introduced into Texas on a grand scale after 1880, the fences closed the once-open range, restricting the movement of all livestock. Sheepmen, coming late to the land, often could not obtain leases for adjoining pastures. It became necessary for them to acquire lots separated by property belonging to others. When moving from one leasehold to another, the sheepmen grazed their animals on pastures owned by cattlemen. The cattlemen objected. To solve the difficulty livestockmen worked out a code that required the herder to drive his flock at least five miles a day on level terrain or at least three miles per day in rougher country. Yet the fences caused a problem, for cowmen found that the sheepmen fenced their ranges, too, restricting the movement of cattle. Cattlemen also fenced their pastures, but unless the fences were of more than three strands, the sheep could walk through them without great difficulty. Furthermore, small farmers and "nesters" pushed into the western livestock regions. Sometimes they found themselves entirely surrounded by a fence without adequate outlet. Or, building their own fences, the farmers cut off sheep and cattle from needed water supplies and grazing land.

Fence-cutting wars followed. Farmers, cattlemen, and herders all cut fences, but secondary accounts support the idea that cattlemen were more aggressive in the art.[20] One concludes that "as early as 1882

[18] Ogden Tanner, *The Ranchers*, p. 120.

[19] Havens, "Sheepmen-Cattlemen Antagonism," p. 19.

[20] See Wayne Gard, "The Fence-Cutters," *Southwestern Historical Quarterly* 51 (July, 1947): 1–15; C. W. Towne and Edward N. Wentworth, *Shepherd's Empire*, 137–39;

. . . if a sheepman . . . fenced his land with anything but rock, it was ten chances to one that his fence would be burned or cut before a year was past." Sometimes even rock fences did not suffice. When Charles Hanna drove the first sheep into Brown County, animosity against the flockmaster was high. To hold the timid creatures at night, Hanna built a rock corral. But, when he arose one morning shortly afterward, he found that gunnysackers had cut the throats of 300 sheep.[21]

Among sheepmen, drifters were the most flagrant fence cutters. Without leases for land, the drifters moved their sheep from range to range, looking for fresh water and lush grass. They cut the fences to get their wagons through and sometimes, especially if the fences contained more than three strands of wire, to enable the sheep to cross into new pastures. A notorious Mexican sheep drifter, hated because of his well-known practice of rolling fence wires to enable his sheep to graze on private property, in 1884 drove his flock into Scurry County east of Snyder, where he met a cowboy. When they got into an argument over watering the sheep at a nearby well, the cowboy shot and seriously wounded the herder. While another drifter in the vicinity hauled the wounded man in a buckboard to a local doctor, the cowboy recruited some of his friends. The cowmen then burned the camp, including the drifter's wagon. Later they hanged the infamous drifter, killed some of his sheep, and scattered the rest.[22]

Cattlemen, using threats and intimidation, tried to keep Guy Mahoney from fencing his land. Mahoney, after buying several thousand head of sheep in 1881, placed the animals on his land in Coleman County. He enclosed the pastures with fence. In 1883 gunnysackers cut his fence for miles, twice between every two posts. While herders repaired the damaged fence, someone placed a full-sized coffin on Mahoney's front porch. Attached to the coffin was a note inscribed, "This will be yours if you keep fencing." Undaunted, Mahoney used the coffin for a water trough.[23]

Mahoney's neighbor and sometime partner Horace Starkweather

Roy D. Holt, "Woes of the Texas Pioneer Sheepman," *Southwestern Sheep and Goat Raiser* 21 (December, 1940): 60–62; Rupert N. Richardson, Ernest Wallace, and Adrian N. Anderson, *Texas, the Lone Star State*, pp. 306–309.

[21] Holt, "Woes of Texas Pioneer Sheepman," pp. 60–62; Havens, "Sheepmen-Cattlemen Antagonism," p. 18.

[22] Havens, "Sheepmen-Cattlemen Antagonism," p. 19.

[23] Holt, "Woes of Texas Pioneer Sheepman," pp. 60–62.

likewise suffered. Wire cutters sliced his fence and burned his sheepfolds, herders' homes, and 2,000 cedar posts piled near his residence. He promptly reconstructed the homes and a five-wire fence with an underpinning of rocks. But someone cut thirty miles of the new fence and turned scab-infested sheep into his flock. This forced him to dip his 8,000 sheep in cold, wintry weather, an act that threatened tragic consequences.[24]

Some sheepmen fought back. C. B. Metcalfe fenced 10,000 acres of the Arden Ranch in Tom Green County. After he had completed four miles of the work, a party of gunnysackers from a ranch north of the sheep spread cut the wire between every two posts. Metcalfe rebuilt the fence and went to town to seek out the owner of the cattle operation. Upon finding the man, the angry Metcalfe, fortified with a loaded shotgun, warned the offending rancher that he would not tolerate additional wire-cutting. Metcalfe was not again molested by fence cutters.

Despite men like Metcalfe, wool growers were at a disadvantage. Not only did they come late to the range, but cowmen outnumbered them. Even the law favored cattle raisers. A Texas law in 1881 (see chapter 7) forbade the grazing of sheep on land belonging to another without the owner's consent. Thus it became necessary, until they found a loophole in the measure, for sheepmen to buy the land on which they ran their sheep, improve it, and protect it by fencing against cattle. When money was invested for fencing, wool growers retained little capital with which to buy flocks. The cattle raiser, however, could put all his money into livestock and graze the sheepman's land, provided it was unfenced. Sheepherders complained bitterly. They protested to the governor, Oran M. Roberts, and the legislature, demanding repeal. Their efforts proved fruitless.

The law made little difference in 1883, when fence-cutting reached epidemic proportions. That year it occurred in more than half of the 171 organized counties. According to Wayne Gard, a severe drought precipitated the war. Water supplies in ponds, wells,and streams disappeared, leaving livestock crazy with thirst. The drought also destroyed the range, leaving pastures bare of grass. To get their livestock to water and food, sheep and cattle ranchers cut fences. The misdemeanor, as Texas law termed it, was most serious along the farmers' frontier,

[24] Ibid.

marked roughly by a line running from near Wichita Falls on the north southward through Brownwood, Mason, and Fredericksburg, to Hondo and Pearsall. In Live Oak County wire cutters, after destroying a fence, dug a grave, dangled a rope in it, and left a sign that said, "This will be your end if you rebuild this fence." In Hamilton County fifteen men, after cutting the barbed wire around a pasture, left a picture of a coffin and a note indicating that they were determined to have free grass and water. At Castroville a farmer found on his fence a card with a bullet hole through it, which read, "If you don't make gates, we will make them for you."[25]

Fence-cutting reached enormous proportions. News of it spread beyond the state. A Chicago newspaper ran an article on the war under the headline, "Hell Breaks Loose in Texas." Local officials and Texas Rangers seemed helpless. The problem got bad enough to cause many sheepmen to collect packs of dogs and ride their fence lines with them each day.[26] To deal with the problem Governor John Ireland called a special session of the legislature. After that body in 1884 made it a felony, fence-cutting slowly disappeared.[27]

Total cost of the fence-cutting wars is difficult to determine. One estimate of the damage placed the loss from destroyed fences at twenty million dollars. The *Fort Worth Gazette* declared that tax valuations declined by more than thirty million dollars as a result of the troubles. In Brown County alone property losses may have exceeded a million dollars. Prospective settlers stayed clear of Texas, and some recently arrived farmers left. Population in many of the Edwards Plateau counties declined.[28]

While fence-cutting declined, other aspects of the sheep wars decreased more slowly. As late as 1890 there was violence. One morning in that year a herder working for W. E. Havens watered his sheep at Big Spring in Howard County. Three cowhands from a nearby cattle operation rode up to challenge the *pastor*. While one of them held his

[25] Gard, "The Fence-Cutters," p. 8.

[26] Ibid., p. 9; Victor I. Pierce, Papers, Southwest Collection, Texas Tech.

[27] Havens, "Livestock and Texas Law," pp. 29–30; Gard, "The Fence-Cutters," p. 11; Edward N. Wentworth, *America's Sheep Trails*, p. 388; Roy D. Holt, "The Saga of Barbed Wire in Tom Green County," *West Texas Historical Association Year Book* 4 (1928): 32–49.

[28] *Fort Worth Gazette*, February 8, 1884; *Galveston News*, January 11, 1884; Gard, "The Fence-Cutters," p. 10.

gun on the sheepman, the others destroyed a dozen of the sheep. Then, as they rode away, each of the cowboys roped an animal and dragged it to death. Many years later T. P. Sloan remembered his father's recalling an incident in 1890 in which cowboys galloped into a sheep camp at night, while the herder slept. The unwelcomed intruders emptied their guns at his pots and pans, scattered his sheep, and sped off before they could be recognized.[29]

In the 1890s, however, the Texas range became a more stable and settled region. Cowboys and herders came to tolerate one another; this in turn led to accommodation and cooperation. In 1896, for example, Walter Posey, later one of Lubbock's most successful bankers, with his brother drove a flock of two thousand sheep from Floydada in Floyd County to Liberal, Kansas. Moving northward they encountered fences on the JA Ranch of Charles Goodnight and John Adair. They entered the JA pastures through a gate and kept the sheep moving at the required three miles a day. When a JA cowboy rode up to question the intruders, Posey indicated that he and his brother were taking the sheep to Kansas. Thereupon the rider warned them about a mountain lion in the vicinity and indicated where the sheepmen might find water, including wells on the JA properties. The cowboy also instructed the young flockmaster how he might secure the horses at night to protect them from the "panther," as the mountain lion was called.[30]

Posey related another example of accommodation. His father, a sheepman in 1896, operated a store in Floydada. Cowboys from such ranches as the XIT, LFD, and Matador did not hesitate to deal with the sheepman. In fact most of them asked the elder Posey to keep on deposit in his store whatever money from their monthly pay they did not spend on personal supplies. Over the years the money accumulated until at one time the Floydada sheepman held $80,000 belonging to cowboys.[31]

By 1900 the sheep wars were over. No compilation of the total damage can be made, but it is clear that threats, intimidation, arson, and murder reached large proportions. While hatred and prejudice

[29] Havens, "Sheepmen-Cattlemen Antagonism," p. 18; T. P. Sloan, interview with Paul H. Carlson, August 23, 1980.

[30] Walter Posey, interview with Seymour V. Connor, July 24, 1956, Southwest Collection, Texas Tech.

[31] Ibid.

lingered for another generation, most western livestockmen by the turn of the century found little difficulty, and considerable advantage, in running both sheep and cattle on their range. By this time, too, many of them, following the lead of Fort McKavett's William L. Black, had added Angora goats to their enclosed pastures.

With the sheep wars over and the Texas livestock triumvirate of sheep, goats, and cattle firmly developed, the Texas range sheep and goat industry at the turn of the century entered a period of reorganization and consolidation. As World War I neared, it approached maturity. The early twentieth century saw the industry change from a speculative proposition to an established institution.

12
The Early Twentieth Century

At the turn of the century in 1900 the Texas sheep and goat industry entered a new age, one marked by increasing sophistication and maturity in all phases of the business. During the period that followed, southwestern wool growers, consolidating their position after the Texas sheep boom peaked, dominated the industry, breathed new life into it, and generally assumed economic leadership. They established commission sales warehouses and organized associations to promote and protect the industry and its members. They struggled to achieve respect and recognition for their industry, but they gained the ideals only slowly before the end of the open range method of sheepherding about 1930.

The history of the sheep and goat industry in the eary twentieth century often seems formless, the events having little logical sequence or relationship to each other. Developments in one area of the industry often had no bearing on events in another. Even so, several changes occurred to alter and refine Texas wool growing. One was that Texas herders acquired a "home on wheels." In 1884 James Candlish, a blacksmith of Rawlins, Wyoming, designed and built the first sheepherders' wagon. Behind his shop he slipped a canvas top on a small wagon, making it look like a shortened prairie schooner. Inside he located at convenient places a stove, water bucket, bed, storage facilities, and lantern. About 1892 a company in Wyoming, using Candlish's model, was annually producing many of the wagons. Its contraption, a compressed bunkhouse, had Dutch doors in front, a cast-iron stove, a bed, compartments for storage, and a canvas or wood covering to block the wind and keep out the cold. Herders, especially on the northern plains readily adopted the sheepwagon. In Texas, because of the generally mild winters and long hot summers, herders were slow to take to the wagon. Because they had to herd their flock back to the wagon each night, some saw it as a nuisance. They preferred a tent and a burro to which they strapped their gear. In this fashion they shifted their campsite with the sheep and continued to sleep on the ground. Nevertheless, the sheepherder's wagon was not an unfamiliar sight after 1900. At lambing and shearing camps several wagons could be found amid the tents and bedrolls on the ground.

Another development was the use of wire-mesh fences—"wolf-proof" fences, as they were called. As late as 1914 sheepmen herded most of the sheep and goats in Texas. With wolves and coyotes numerous everywhere in the Southwest, wool growers sustained heavy losses whenever sheep or goats strayed from the flock or ranchers allowed them to remain out over night. As the number of herders declined, forcing wages too high to make some operations profitable, farmers and ranchers looked to other means of handling small livestock. The net-wire fence proved an answer. And just as barbed-wire fences in the late 1870s and afterward effectively ended the era of picturesque and romantic cowboys, wolf-proof fences eventually brought an end to the old-time *pastor*. It was a slow process, however, as many Texas sheepmen, reluctant to spend enormous sums to fence their pastures, turned to Europe for herders. Basques from the high mountain province in Spain arrived in increasing numbers to take over from Englishmen, Germans, and even many Mexicans the responsibility of herding sheep.[1]

But net-wire fences eventually replaced all herders. Even before 1900, sheepmen industriously experimented with barbed-wire fences as a possible solution. Texans, following the examples of farmers in Illinois and cattlemen everywhere, took the lead in the West. Colonel Black enclosed some 30,000 acres on the headwaters of the San Saba River near Fort McKavett. Bob Wylie fenced a large pasture on the Colorado River above Ballinger. John A. Loomis enclosed his pasture on the Kickapoo in Concho County. As a unique touch he added, according to at least one account, an electric signal system so that the opening of a gate anywhere on his property was recorded at the ranch.[2]

[1]Ray Willoughby, interview with Paul H. Carlson, April 28, 1977, Southwest Collection, Texas Tech University, Lubbock; *Wall Street Journal*, April 14, 1978.

[2]"New Era Dawns with Net-Wire Fences," *Sheep and Goat Raiser* 50 (October, 1969): 30, 38.

But the sheepmen found that barbed-wire fences left much to be desired. Although they enclosed the land, giving the owner exclusive use of it, barbed-wire fences did not prevent predators from destroying his sheep or lessen the need for herders, since sheep could often walk through fences of only three strands. Wool growers needed a different fence. They found one in the mesh-wire variety, called by one historian "the most important improvement in the sheep industry in Texas."[3]

There is considerable controversy concerning the invention of woven-wire fences. One authority states that mesh-wire fences originated in Australia before they became the pattern for the United States. Another account states that the woven-wire fence first appeared in 1883, the invention of an imaginative man who lived on a farm in Michigan. A third account gives credit for designing the first woven-wire fence, as well as a wire-weaving machine, to Peter Sommer of Tremont, Illinois. Others improved on his work, first called the Keystone fence, during the years Theodore Roosevelt served as president of the United States and accordingly renamed it the Square Deal fence.[4]

Among sheepmen in Texas, Sam Hill of Tierra Alta, near San Angelo in Tom Green County, built one of the first wolf-proof fences. Born in 1859, Hill herded sheep as a youth near Austin. In 1880 he began raising sheep for himself. Later, at the turn of the century, he began running cattle and goats in Tom Green County. He also began a vigorous fight against predatory animals in his immediate vicinity. Short and stocky, Hill was a man of tremendous energy, who for more than ten years was the unofficial ambassador of the Texas sheep and goat industry in Washington, D.C., and elsewhere. His fence, which created a profound impression among sheepmen in West Texas, initiated a boom in woven-wire fences.[5]

By 1910 many farmers and ranchers had come to recognize wolf-

[3]Harold M. Gober, "The Sheep Industry in Sterling County," *West Texas Historical Association Year Book* 27 (1951): 47.

[4]"Wolf Proof Fences in Texas," *Cattleman* 18 (September, 1931): 27; "New Era Dawns," p. 38.

[5]"Sam Hill, Leader for Years, Dies at Tierra Alta," *Sheep and Goat Raisers' Magazine* 13 (July, 1933): 125; John P. Classen, "History of Predatory Animal Work of the Bureau of Biological Survey in Texas," *Sheep and Goat Raisers' Magazine* 11 (June, 1931): 295; "Sam Hill, Pioneer Builder of West Texas," *Sheep and Goat Raisers' Magazine* 7 (September, 1926): 8.

proof fences as essential. In Sterling County Charles M. Crawford, who netted his ranch in 1914, became the first sheepman to put up wolf-proof fencing. He claimed that the netted pasture helped to keep out predatory animals and contributed to the increase and improvement of the production of lambs and wool. T. A. Kincaid of Ozona in 1915 built the first mesh-wire fence in Crockett County. Later he added more property by lease and purchase, bringing his holdings to seventy sections. On this he erected 175 miles of mesh-wire fence in thirty-two pastures and traps. Without herders, he grazed nearly 15,000 sheep, more than 3,000 goats, and 1,000 cattle. He sheared between 90,000 and 110,000 pounds of wool every year. Not until the 1920s, however, did most Texas sheepmen erect woven-wire fences. Then the practice spread quickly. At the end of the decade one authority conservatively estimated that 95 percent of the state's sheep and goats grazed in net-wire pastures.[6]

As time passed, the fences came to conform to a common pattern. In the early twentieth century the woven wire, with a six-inch mesh and standing from forty-two to fifty-two inches in height, was stretched on cedar posts. Builders set barbed wire along the ground, sometimes on both sides of the posts, to prevent wolves and coyotes from scratching under the fence, and frequently placed three barbed wires above the woven wire. The standard wolf-proof fence cost about $650 per mile to buy and erect.

There were several advantages for "loose" grazing the flocks within these enclosed pastures. J. M. Jones of the Texas Agricultural Experiment Station in 1936 noted that such grazing reduced handling costs, increased the grassland's carrying capacity, resulted in fewer communicable-disease problems, produced heavier and cleaner fleeces, and assisted in obtaining a larger percentage lamb crop, which weighed more at weaning time.[7]

The turn of the century also found ranchers elsewhere in the Southwest adopting the Texas livestock triumvirate of cattle, sheep,

[6] W. G. Rawls, San Angelo, letter to Harold M. Gober, May 31, 1950, in Gober, "Sheep Industry in Sterling County," p. 49; "T. A. Kincaid's Contribution to the Sheep and Goat Industry," *Sheep and Goat Raiser* 30 (Jaunary, 1950): 20–23; J. M. Jones, "History of the Range Sheep Industry in Texas," *Southwestern Sheep and Goat Raiser* 6 (March, 1936): 32.

[7] Jones, "History of Range Sheep Industry," p. 32.

and goats. Spaniards had done this in Texas for centuries. Anglo ranchers had done it before the Civil War, but in the years afterward, with exceptions in the Hill Country, only a few men, such as Richard King, William Black, and Milton Favor, grazed sheep with cattle. Few Anglo-Americans were willing to risk the social stigma of "fooling with sheep." When such foreigners, promoters, or entrepreneurs as Black not only raised sheep, goats, and cattle on the same range, but also bragged about their success, some other ranchers followed suit. But there was no general acceptance of the practice. In Sterling County, for example, in 1886, when there were plenty of cattle, only a few settlers had sheep. In the early 1890s the number of sheep declined, but it increased after 1893, as more ranchers accepted the animals. By 1920 practically every livestockman in the county grazed both sheep and cattle. Most also browsed goats. In 1951 only one local cattleman did not maintain sheep.[8] Ranchers in the Trans-Pecos from the 1890s or before raised both sheep and cattle, and in the early twentieth century the practice became standard on the Edwards Plateau.

As Texans experimented with cattle, sheep, and goats on the same range, large-scale commercial sheep feeding caught on. Sheep feeding—raising sheep in pens or fenced yards on foodstuffs rather than grazing them on fresh pastureland—had always existed in America, but not until after the Civil War did it reach a substantial scale, and not until after 1900 did it assume major importance. The movement began in the East before high land prices forced a shift to the western Plains states. Nebraska, Kansas, and Colorado led, followed by Montana, Wyoming, the Dakotas, and Minnesota. Sheep feeders fattened wethers and ewes for winter markets in Chicago, Fort Worth, Saint Paul, and Kansas City slaughtering plants.[9] After 1920 there was a marked increase in lamb feeding in Missouri and northeastern Iowa. For at least two decades after World War I Texans shipped their lambs northward to graze on winter wheat fields in Kansas and Oklahoma. The fat ones they sorted out in December and early January for slaughter, while those still unfinished they moved to drylots in Kansas, Nebraska, and

[8] Gober, "Sheep Industry in Sterling County," p. 43.

[9] U.S. Department of Agriculture, *Livestock on Farms, January 11, 1867–1919*, Bulletin of the Bureau of Agricultural Economics, 1938, p. 27; U.S. Department of Agriculture, *Livestock, Meats, and Wool—Market Statistics and Related Data*, Bulletin of the Bureau of Agricultural Economics, 1942, p. 10.

Colorado. Feeding in Texas became important on the High Plains in Lubbock and surrounding counties and near Amarillo, but only at the close of the open range. Subsequently, Texas sheepmen fattened their animals in several areas, but especially in various sections of the state between Fort Worth and San Angelo.[10]

In Texas feeding operations before 1930, the fed-lamb period ran from mid-November into late May. Most operators used self-feeder hayracks. They fed their animals in the open, where the air was bracing, the rainfall light, the snow usually dry, and a large portion of the winters sunny. Where there was a danger of driving storms, if no other shelter was available, the operators constructed windbreaks. The location of feeding operations shifted with changes in land prices, relative costs, transportation conditions, and availability of foodstuffs for rations. The development of irrigation, the increased production of alfalfa and barley, and the availability of other feed, especially corn, made possible the growth of commercial feeding. The number of sheep and lambs on feed remained small until the mid-1920s, when dramatic growth occurred. In 1927 Texas counted 23,000 animals on feed. The next year the number increased to 68,000, and by 1930 the state had reached a total of some 83,000.[11]

World War I had important consequences for the Texas sheep and goat industry. The war produced at least two significant trends: a raise-more-sheep campaign and the fixing of wool prices. The raise-more-sheep campaign started as part of the preparedness movement that swept the country following the outbreak of war in Europe in August, 1914. It reached a peak in the spring of 1918, when President Woodrow Wilson established a small flock of sheep on the White House lawn—symbolic of the country's desire to increase wool and lamb production—and it ended rather abruptly in early 1919, as peace negotiations began in Paris and domestic wartime agencies, such as the powerful War Industries Board and the Food Administration, discontinued their operations. During the war, retail meat-market operators over much of the eastern part of the United States placed signs in their windows proclaiming "Eat No Lamb" and "Eat No Veal." Hotels, restaurants, and even railroads, believing that such a boycott would mean more sheep

[10] Edward N. Wentworth, *America's Sheep Trails*, p. 377.

[11] Texas Crop and Livestock Reporting Service, *Texas Historic Livestock Statistics, 1867–1976*, pp. 27–28.

and wool, eliminated lamb from menu cards. The boycott, however, tended to operate in reverse, lowering prices Texas and other sheepmen received for their animals and thus discouraging ranchers from expanding their operations.[12]

Nevertheless, at the end of 1917 the raise-more-sheep campaign was in full swing. The fighting in Europe, creating a demand from allies for food, had caused a tremendous slaughter of meat animals and had helped lead to the campaign. American producers, including sheep and goat ranchers, had to supply their own soldiers and citizens as well as help supply the Allies. In a government proclamation every sheep raiser was urged, for breeding purposes, to "carry to its yearling form every ewe lamb that promises to have an economic future."[13] Some leaders of the raise-more-sheep campaign asked President Wilson for legislation that would force farmers to increase their sheep production. Although no such measure passed Congress, the drive for sheep expansion continued through most of the next year.

In the early fall of 1918 the campaign noticeably slackened. Lamb and mutton prices, which had peaked out in the spring and summer, began to decline. A shortage of railroad stockcars in the Midwest helped to dramatize the price decline. With the lack of cars causing a transportation tie-up along the Missouri River, shippers unloaded large numbers of sheep and lambs, carried over western railroads, at all river feeding and unloading points, resulting in an oversupply of sheep at markets in nearby Omaha and Kansas City. In Omaha shippers marketed as many as 65,000 head in one day, causing a break in prices of up to $1.50. The drop in price, although not as drastic elsewhere, spread across the nation, affecting Texas two days later. John Ashner of Tom Green County reported that he got $1.25 per head less than he had expected.[14]

With the end of the war in November, the raise-more-sheep campaign floundered. The armistice on the eleventh reduced the urgency for more meat. Ranchers were disgruntled. Many of them had bought a few sheep in response to the raise-more-sheep campaign, but when

[12]Cited in "World War I's Impact on Wool and Lamb," *National Wool Grower* 55 (January, 1965): 47.

[13]Ibid.

[14]Newspaper clipping, Morgue Files, Texas Sheep and Goat Raisers Association, Records, Southwest Collection, Texas Tech.

they lost money on the proposition and the end of the war reduced the need for sheep, they complained that the campaign was mere propaganda that had got them unwittingly into an unsuccessful enterprise. In addition the National Wool Growers Association feared that the eastern lamb and mutton boycott might change eating habits by turning Americans away from lamb in favor of beef or pork. To counter such a possibility, it launched an "Eat-More-Lamb" campaign. The raise-more-sheep program slowed in late 1918 and died out early the next year.

Because it gave new importance to sheep raisers and helped the country meet its wartime food demands, the campaign proved effective. The number of stock sheep and lambs on farms in the United States had declined from 40.5 million in 1913 to 35.2 million during the first year of the war. As the raise-more-sheep campaign took effect, the number of animals on farms increased again, reaching 38.4 million in 1919.[15] Texas experienced a fluctuation similar to the larger national one. The number of stock sheep dropped from 2.3 million in 1912 to 2.2 million in 1917. It reached 2.6 million two years later.[16]

Similarly, the fixing of wool prices during World War I was a matter of importance. In Boston, the major wool warehousing, buying, and handling center in the United States, the prices in 1914 on wool in grease (unwashed) was 22.8¢ per pound. The demands of the war, including large manufacturing contracts for outfitting the army and navy, drove the price rapidly upward to 59.6¢ per pound in 1917 and 69.6¢ a year later. Prices on clean wool were even higher, ranging in the summer of 1917 from $1.06 for inferior types of common and braid wools to $1.68 for choice half-blood staple, a remarkable rise over the 14¢ per pound growers had earned in the mid-1890s.[17]

The sudden jump in wool prices did not go unnoticed in Washington. Bernard M. Baruch, head of the War Industries Board and chairman of the raw materials section of the Council of Defense, appointed a committee of six members, three wool growers and three wool manufacturers, to study all phases of the wool industry as it applied to the

[15] T. R. Hamilton, "Trends in the Sheep Industry in Texas," *Sheep and Goat Raiser* 25 (June, 1945): 7.

[16] Texas Crop and Livestock Reporting Service, *Texas Historic Livestock Statistics*, p. 27.

[17] Ibid.; "World War I's Impact," pp. 49–50.

war. Among other things, the committee considered the fixing of wool prices, but it recommended to an advisory committee of the Council of Defense against such an action. Nevertheless, as prices for wool continued to rise, the government, through the Council of Defense, in April, 1918, established maximum prices it would pay for wool and created a wool division within the War Industries Board. Lewis Penwell of Montana became chief of the wool division, and Charles J. Nichols was named wool administrator. The War Industries Board set prices at those prevailing in July, 1917, or from four to ten cents per pound below the 1918 level. Wool manufacturers applauded the step; the wool growers believed the action was unfair but accepted it when the government agreed to assume the cost of handling wool after owners had delivered it to local dealers.

On the whole, the price-fixing plan worked smoothly. Some sheep-industry leaders hoped it would lead to changes in wool marketing procedures, with more consignments and less speculation, but these effects did not last. When signs of peace appeared in the fall of 1918, wool prices fell again, dropping below the maximum levels the government had been paying. When Washington officials terminated the War Industries Board in January, 1919, all hope for continuation of the government wool program faded, and the producing, marketing, and delivering of wool reverted quickly to the prewar status.

Those developments, thanks to the state's wool and mohair warehouse system, did not adversely affect Texas. T. C. Frost in San Antonio and Charles Schreiner in Kerrville had established the first warehouse commission operations. Others had soon followed them, but at the end of the nineteenth century most wool and mohair producers in Texas continued to sell to eastern buyers, who moved about the state with little system. Growers found themselves at a considerable disadvantage and seldom got more than the price buyers offered. The weakest part of the growers' position was that their wool and mohair had to be sold; they had no safe place to hold it. Furthermore, growers were often poorly informed on the relative values of the various grades of wool or mohair. They had no source of information on market condition. In other words, they were almost wholly at the mercy of the buyer.

The Frost and Schreiner operations proved a boon for Texas wool growers, but by 1900 Frost had left the business, and Kerrville was far away from the main centers of sheep and goat production. To consider improvements in wool marketing several men met in 1908 at Ozona. They discussed plans to form an organization designed to buy, store, and sell wool and mohair. Out of their meeting emerged the Wool Growers Central Storage Company of San Angelo. Leaders in the promotion of the warehouse operation included Robert Massie, S. E. Couch, Colonel C. C. Walsh, and W. B. Sayers. Capitalized at $100,000 and with offices and warehouse in San Angelo, the Wool Growers Central Storage Company was the first such entity incorporated in Texas. Massie, a Crockett County rancher and San Angelo businessman, became the president and held that office for many years afterward. An immediate success, the company built additional warehouses, increased its capitalization, offered a valuable service to Texas sheep and goat raisers, and became one of the biggest wool- and mohair-concentration houses in the United States.

Following the organization of the Wool Growers Central Storage Company, Texas businessmen and ranchers established additional warehouses from time to time. By 1928 a score or more existed in the West Texas wool and mohair regions. Among these were three in Del Rio, two in Rocksprings, two in Menard, two in Uvalde, another in San Angelo, and others in several other cities and towns. They all operated along the same general lines. That is, when warehouse operators held sales, representatives from the leading wool and mohair mills of the country attended to deal with a sales committee composed of wool growers and wool and mohair classification experts. The buyers offered a price for a particular clip. If the committee accepted the offer, a sale was consummated. If not, the committee held the clip to offer it for sale later. The warehouse usually retained a 2½ percent commission from the sale of all wool and mohair.[18]

In another significant development, ranchers created the American Angora Goat Breeders Association, established in 1898 at Kansas City, Missouri. Members moved its headquarters to Sabinal, Texas, in 1924 and two years later to Rocksprings in Edwards County, where it

[18] S. P. Shaw, "Development of Wool and Mohair Warehouses in Texas," *Sheep and Goat Raisers' Magazine* 8 (May, 1928): 67–70.

remains today. The association keeps records showing pedigrees of all purebred Angora goats and issues certificates of registration on each animal. It advertises the industry, offers premiums to Angora goat shows, and otherwise maintains records for Angora goat breeders.[19]

Similarly, ranchers created the Texas Sheep and Goat Raisers Association. In January, 1913, several men at Sonora organized the Stockman's Protective Association of Sutton County. Their object was to fight thieves who stole sheep and goats. Assessing a small levy on each of some forty-five members, the association hired an inspector, put him to work, and carried on the program for two years with considerable success. Because of their success, some members of the association desired to create a statewide organization. Thus, in Del Rio on January 13, 1915, J. B. Murrah, V. A. Brown, Johnson Robertson, E. E. Stricklin, and B. M. Halbert, all goatmen, met with some thirty others. They had two goals in mind: to form a state association of sheep and goat raisers for their mutual advancement and protection and to ask the Texas legislature for an appropriation to establish a livestock experiment station in West Texas. While it took two years to get the experiment station, located between Sonora and Rocksprings, the association was formed before the day was out. J. B. Murrah of Del Rio became the first president. Subsequent presidents included B. M. Halbert, J. B. Moore, Bob Martin, Claude Broome, E. K. Fawcett, and T. A. Kincaid, Sr., who served for eleven years. In 1935 the group merged with the Texas Wool and Mohair Growers Association, organized in 1933, and the Texas Warehouse Association. The new structure established headquarters in San Angelo, with Abe Mayer as its president.[20]

The Texas Sheep and Goat Raisers Association has promoted the industry. It has successfully sought state and national legislation protecting Texas ranchers, waged an endless campaign against predators, sponsored wool and mohair fairs, and cooperated with such national organizations as the Mohair Council of America and the National Wool

[19]See American Angora Goat Breeders Association, Records, Southwest Collection, Texas Tech.

[20]"History of the Texas Sheep and Goat Raisers Association" (typescript), "V. A. Brown" (typescript), and B. M. Halbert, "The Organization of the Texas Sheep and Goat Raisers Association" (newspaper clipping), all in Morgue Files, Texas Sheep and Goat Raisers Association, Records, Southwest Collection, Texas Tech.

Growers Association. It holds an annual meeting at which hundreds of Texas ranchers meet to exchange ideas and to discuss common problems.

In 1920 the Texas Sheep and Goat Raisers Association published the first issue of the *Sheep and Goat Raisers' Magazine*. The publication was designed to inform association members about activities of the organization. Over the years the periodical has undergone several name changes, and it has absorbed some related journals, such as the *Angora Journal*. Today the publication is entitled the *Ranch Magazine*. It has also undergone changes in style and format. For a time in the 1930s and early 1940s it appeared twice each month, but today it is a monthly magazine. It has been an important organ through which Texas sheep and goat raisers can get better acquainted with their industry and study their business intelligently.

As these developments occurred, the old range sheep and goat industry in Texas came to an end. With its close, the industry changed, becoming a secure agricultural enterprise. In the process, the species of early sheepmen became almost extinct. The poorly educated, sometimes erratic genius, who made thousands or even millions of dollars through native shrewdness, flashes of intuition, and perseverance, was well equipped to build a sheep empire but not to administer it. With herders in short supply and pastures netted with wolf-proof fences, the new species of sheepmen took over the ranchland. Modern successful wool growers added cattle and goats, scientifically bred their animals, and applied available technology to their enterprises. They placed greater emphasis upon education, studied the wool and mohair manufacturing branches of the industry, and adopted sound business methods for operating their ranches.

By this time Texas had established itself among the leading sheep- and wool-producing states in the nation. In doing so, the state had enjoyed a pattern of growth that differed from the pattern of growth for the United States. From 1870 to World War II sheep production in the United States showed six definite cycles. The cycles lasted from seven to fourteen years, averaging ten years, as shown in table 7.

The cycle movement was not duplicated in Texas for this period. From 1870 to World War II sheep numbers in the state showed an upward trend, with the exception of breaks in 1886, 1913, 1923, and 1934. In 1943 Texas reached a peak in numbers of sheep and lambs, with

TABLE 7
Sheep Production Cycles in the United States, 1870–1942

Years, peak to peak	Length of cycle in U.S. sheep production (in years)
1870–84	14
1884–93	9
1893–1902	9
1902–1909	7
1909–19	10
1919–32	13
1932–42	10

SOURCE: T. R. Hamilton, "Trends in the Sheep Industry in Texas," *Sheep and Goat Raiser* 25 (June, 1945): 6.

more than 10,800,000; afterward, there occurred a slow but almost steady decline. During the period of growth, however, Texas produced an increasing proportion of the total sheep and wool of the United States. In 1910 Texas produced only 4 percent of the total number of sheep in the United States, but in 1920 the figure stood at 9 percent. Ten years later, when the state had become the nation's leading producer, Texas accounted for 13.8 percent of the sheep and 13.7 percent of the wool output of the country. Greater growth occurred in the 1930s, when Texas claimed more than a fifth of the sheep and wool of the United States. The state has maintained that percentage to the present time.[21]

Angora goat and mohair producton trends differed from those of sheep and wool. Before 1900 Texas not only assumed the lead among states in production of goats and mohair but also far outdistanced the others. As early as 1910, Texas ranchers produced 71.2 percent of the mohair clip and nearly as large a percentage of goats. In 1920 the Texas mohair clip represented 79.2 percent of the total United States production. Ten years later the figure was 84.2 percent. Today Texas goatmen raise 90 percent of the country's Angora goats and nearly 97 percent of its mohair. Because the state produced such a large percentage of the

[21] Texas Crop and Livestock Reporting Service, *Texas Historic Livestock Statistics*, pp. 27–28.

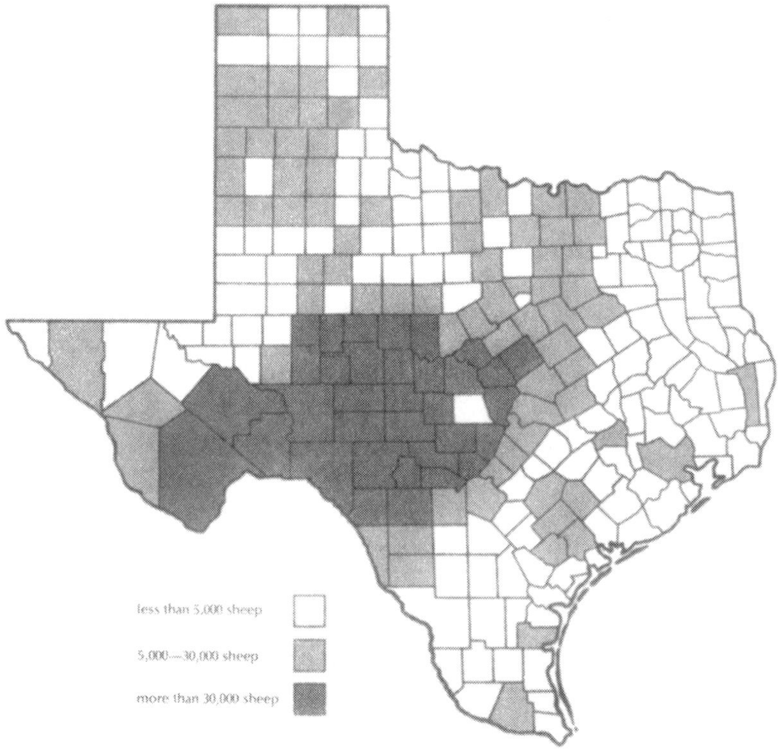

Distribution of sheep in Texas, by county, 1930

country's total, the year-to-year production changes for Texas checked closely with year-to-year changes in the United States. In Texas from 1900 to 1942, with the exception of breaks in 1922, 1932, and 1942, rapid growth characterized production of mohair. This steady upward trend was not so evident in the production figures of other states; a cyclical movement characterized them rather than a trend. The period of most rapid growth came after 1922, as Texas contributed an increasing proportion of the mohair clipped in America. Indeed, for nearly half a century after the range method of goat raising ended, Texas produced more mohair than either Turkey or South Africa, the next two leading countries after the United States in the production of mohair.[22]

[22]T. R. Hamilton, "Trends in Production, Use and Prices of Texas Mohair," *Sheep and Goat Raiser Magazine* 25 (July, 1945): 14.

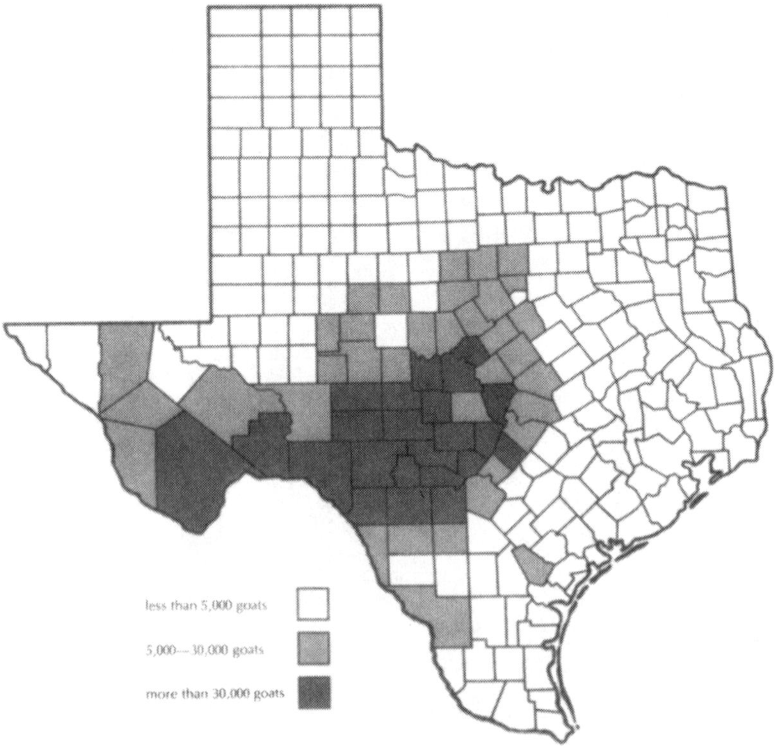

Distribution of goats in Texas, by county, 1930

About 1930, at the end of the range method of sheep and goat raising, after ranchers had widely adopted loose herding within mesh-wire fences, Texas sheepmen could look back with considerable pride at what they had accomplished. Farmers and ranchers in the state had made several contributions to the range sheep and goat industry. They had maintained with modifications the colorful Spanish system of herding and ranch management and had helped to spread those practices to several states in the Southwest. More than any other rancher in America, George Wilkins Kendall, the successful Hill Country sheepman, had promoted western wool growing, and his methods of dipping sheep against scab and other diseases had attracted enough attention to be widely copied. Eventually his methods became the standard practice for treating sheep against a variety of parasites. The wool ware-

TABLE 8
Three Leading States in the U.S. Production of Sheep and Wool, 1859–1943

Year	Rank in Sheep and Wool Production		
	1	2	3
1859	Utah	Kansas	Ohio
1869	Ohio	California	New York
1879	Ohio	California	Michigan
1889	California	Ohio	Texas
1899	Montana	Wyoming	Ohio
1909	Wyoming	Montana	Ohio
1919	Wyoming	Montana	Idaho
1929	Texas	Montana	Wyoming
1939	Texas	Wyoming	California
1943	Texas	Wyoming	Montana

SOURCE: T. R. Hamilton, "Trends in the Sheep Industry in Texas," *Sheep and Goat Raiser* 25 (June, 1945): 6.

TABLE 9
Four Leading States in the U.S. Production of Angora Goats and Mohair, 1900–50

Year	Rank in Goat and Mohair Production			
	1	2	3	4
1900	Texas	Oregon	California	Iowa
1910	Texas	Oregon	New Mexico	Arizona
1920	Texas	Arizona	New Mexico	Oregon
1930	Texas	Arizona	New Mexico	Oregon
1940	Texas	Arizona	New Mexico	Oregon
1950	Texas	Oregon	Missouri	New Mexico

SOURCE: H. A. Mauersberger, ed., *Mathew's Textile Fibers*, pp. 2–37, 678.

house system Texans had established proved to be a unique livestock-marketing contribution, for no other state developed such widespread use of warehousing for wool and mohair. By the close of the range sheep industry there existed well over one hundred wool and mohair warehouses across the state, handling about 90 percent of the Texas clip. Herding within wolf-proof fences was another highly distinguishing feature of the Texas livestock industry. In the early days Texas was the only western sheep state in which ranchers followed the practice.

Clearly, the Texas range sheep and goat industry to 1930 must be judged as one of the state's most remarkable enterprises. Through more than two centuries it had a positive effect on the Texas way of life. It provided lean, red meat for food and enormous quantities of relatively cheap fiber for clothes, furniture covers, draperies, and other items. Economically, it provided jobs in growing, transporting, processing, and marketing its products. When pioneers were pushing westward, it offered a stable enterprise to settlers seeking new homes on the Texas frontier. Culturally, it contributed an abundance of new words—*cabrito*, for example, and terms such as wolf-proof fence—to the English language. It also gave the state new institutions, such as the sheepwagon and wool warehousing. Socially, its impact was substantial. The industry encouraged immigration of Mexicans and Basques. It gave employment to penniless English, German, and Scottish immigrants. Through various professional organizations, it promoted fraternal and economic associations for its members, published magazines and journals, championed hybrid animals, rural improvements, and a large number of other items that have improved the quality of life in the state. The Texas range sheep and goat industry has earned a respected place in the life and lore of the state.

Epiologue
Texas Range Sheep and Goat Industry since 1980

Not surprisingly, since *Texas Woollybacks* first appeared in 1982, the Texas sheep and goat industry has evolved and enjoyed some successes but faced many familiar challenges. Costs for equipment and feed, for example, have risen over the years, while prices received by producers of wool, mohair, goat, lamb, and fat sheep have remained, although higher, relatively stable. Marketing schemes, like the fate of older ones such as "Miss Wool of America Pageant," "lamberger," and "chevon," have met with mixed results. The industry continues to face difficult issues like foreign competition, high labor costs, declining markets, and severe, prolonged droughts.

Nonetheless, raising sheep and goats in Texas after 1980 remained a viable economic endeavor. Ranchers paid greater attention to range land conditions, managed their flocks carefully, checked market swings, sought additional income through hunting leases and related schemes, and experimented with new breeds and a large variety of crosses. As a result, in the late twentieth century, sheep and goat cross breeds proliferated. The Spanish Merino, the foundation breed for all fine wool sheep, along with its various crosses—especially the Delaine-Merino—still continued to rule the wool trade in the twenty-first century; however, the Rambouillet, a French offshoot of the Merino, also remained popular in Texas.

Additional breeds, although around for a century, gained popularity, particularly Southdown, Hampshire, and Suffolk pure breeds and mixes. Southdowns, one of the oldest breeds of sheep, are small to medium in size with gray to light brown faces and wool on their legs. They are best suited to farm flock production. Hampshires, popular for both meat and wool, are large in size with black faces and wool on the legs. They are adaptable to varied climates and fast-growing with medium, easy-to-spin wool. Suffolks are large sheep with bare heads, black faces, and bare, black legs. They have a strong milking and lambing ability, adapt well to heat and cold, and produce high quality meat carcasses and medium, easy-to-spin wool.[1]

Some ranchers turned to lamb and mutton production. As a result, so-called "hair breeds" such as Dorper, Katahdin, and Royal White gained acceptance. Dorpers, developed in South Africa, are primarily mutton sheep. Very hardy, they thrive under range conditions where other breeds struggle to exist. Strong and non-selective grazers, they have been incorporated in Texas into well-planned range management systems. Katahdin, developed in the United States, are hardy, adaptable, and low maintenance animals with strong flocking instincts. They do not require shearing, but they do produce a thick winter coat that sheds in the spring. Royal Whites, developed near Hermleigh, Texas, by William Hoag, are pure white and shed their hair in the spring. They require little human support and thus produce some labor benefits. The female Royal Whites have a strong mothering instinct and produce large crops of lambs that survive to weaning.[2]

In early 2011, Texans raised some 880,000 sheep and lambs valued at $113.5 million. While much improved over 2010 figures, the number of sheep and lambs had dropped considerably from 1943, when state growers produced 10,830,000 sheep and lambs, and from 1973 numbers of 3,210,000. Likewise, wool production decreased from 26,400,000 pounds, valued at $23.2 million, in 1973 to 3,600,000 pounds, valued at $5.74 million, in 2010.[3] Numbers and values of sheep have always fluctuated, of course, but a downward trend characterized the period after 1980. Cash receipts for sheep and lambs in 2000, for example, stood at $38.2 million. In 2005, they reached $53.35 million, but they fell again in 2009 to $37.9 million. The same trend occurred with wool; cash receipts in 2000 amounted to $3.6 million, increased to $5.3 million in 2005, then fell to $3.6 million four years later.[4]

In 2014, most sheep, lambs, and wool in Texas came from the Edwards Plateau area of West Central Texas and nearby counties. The leading counties included Concho, Crockett, Edwards, Gillespie, Mills, Pecos, Schleicher, Sutton, Tom Green, and Val Verde. San Angelo remained the largest sheep and wool market in both Texas and the nation and is a center for wool and mohair warehouses, scouring plants, and slaughterhouses. Among states, Texas continued to be the largest producer of sheep, lambs, and wool.[5]

Like sheep and wool producers, goat raisers also shifted breeds in response to the American marketplace. Angora goats with their long and versatile mohair fleece remained popular, but their popularity declined. In 2010, Angoras on Texas farms and ranches numbered only 120,000 animals out of a total goat population in the state of approximately 1,080,000. That year, mohair from clipped goats reached about $3.07 million in value.[6]

Despite the declining Angora and mohair numbers, Texas, as it has been since the early twentieth century, maintained its status as a global competitor. Indeed, according a recent *Texas Almanac*, in 2015 the state produced nearly half the world's supply of mohair and eighty-four percent of the United States clip. Although producers raised sheep and goats in many parts of the state, the Edwards Plateau and Texas Hill Country led the industry. The leading counties for Angora and mohair production included Crockett, Edwards, Gillespie, Kendall, Schleicher, and Val Verde. San Antonio and Laredo were the chief commercial centers for mohair with San Antonio dominating the trade, but Rocksprings, in Edwards County, served as the unofficial capital of the Texas Angora goat and mohair industry.[7]

After 1980, there was an increased emphasis on meat and dairy goats. Among them, boar goats, first brought to the United States in 1993, represented perhaps the most popular of the meat animals. They are large in size with short horns and large lop ears, and they show a variety of color combinations. Ranchers crossed them with Angoras, dairy goats, and other breeds, both purebloods and crosses, to improve meat production.

Fainting (myotonic) goats go by any number of names, including "scare goats." First developed in the United States, they have a condition called myotonia congenital, in which muscle cells cause an exaggerated response to such stimuli as a bump or sharp noise. Because their muscles do not relax quickly after contraction, they may fall down and appear to faint. The condition may last fifteen seconds. Despite the defect, they are exceptional meat animals.

Spanish goats, or at least modern crosses of them, are still around—although they are almost unrecognizable from their early forms when they first arrived with Spaniards in Texas and New Mexico. They are primarily meat animals, but ranchers successfully breed them to boar goats, dairy goats, Angoras, and others.

Dairy goats have been described as hardy, gentle, intelligent animals. Like most goats, they are browsers who prefer brush lands to pastures, and they will eat non-noxious weeds and other plants. Their milk averages 3.5 percent butterfat. In 2010 and later, goat raisers throughout the state received $108.2 in cash receipts for the sale of meat, dairy, and Angora goat products.[8]

In the late twentieth and early twenty-first centuries, sheep and goat raisers continued to face legislative challenges. On the plus side, in 1986 their representatives in the United States Congress pushed through an extension of the National Wool Act. First passed in 1954 and signed into law by President Dwight D. Eisenhower, the law, in absence of tariff protection, provided ranchers with incentive payments for every pound of wool they produced. Later, the government added mohair to the program. The Department of Agriculture determined the incentive level and over the years the level tended to increase, thus providing important financial help to wool growers.[9]

Legislation pertaining to water and water rights also concerned sheep and goat producers. In 2000, the *Livestock Weekly* carried an article titled "Edwards Aquifer Head Reviews Groundwater Status in Texas." The problem, as smaller wool and mohair growers in the region saw it, was that under Texas water law, big companies—industrial firms, or, less often, large ranchers—could and did suck shallow wells on their neighbor's property dry. The "rule of capture" or "right of capture" is a tort law in Texas that says, according to the article, "you have no duty or care to your neighbor for the operation of a well on your land." [10] Clearly, water shortages remained a problem. In 2012, state legislators created a Texas water plan aimed at ameliorating terribly dry conditions from a long drought, but there seemed to be no political will to implement the plan. Partly as a result, water, or lack of it, persisted as a significant ranch problem. In addition, polluted streams, brush as major water waster, declining water tables, and dry stream beds compounded the Texas sheep and goat raisers problems.

Sheep and goat raisers opposed some legislation. In 2012, the Texas legislature considered a law to regulate the amount of dust allowed from sheep-feeding operations—whether the dust came from a feed yard or a ranch. The industry wanted no part of the bill. Additionally, the federal government planned to eliminate the United States Department of Wildlife Service, an agency that provided assistance to ranchers who raised imported (or exotic) wild animals.

Indeed, the raising of exotics, as they were called, was an important development in late twentieth century West Texas. Some ranchers added exotics to their grazing lands while they continued to produce cattle, sheep, or goats. Others eliminated traditional livestock and turned only to exotics—both breeding and raising them—for hunting purposes. Land owners built eight- to ten-foot high fences and at good profit leased their ranges or portions of them to big-game hunters. The exotics included African antelope, addax, blackbuck, gazelle, impala, and zebra, among others. The exotic industry in Texas in 2012 was a $1.3 billion industry responsible for 14,000 jobs in the state.[11]

Relatedly, the number of South American vicunas, alpacas, and lamas increased. Vicunas, as a feral cousin to alpacas and lamas, should perhaps be considered an exotic. Nonetheless, growers sought each of the high mountain species for their wool. In 2011, raisers in the United States registered 150,000 pure bred alpacas, most of them in Texas. Adult alpacas provided on average of ten to eleven pounds of wool.

In the early twentieth-first century, some sheep and wool producers—the numbers in 2015 were increasing—began raising bison. It was a common trend on large ranches located across the American Great Plains. Bison, or buffalo, meat is lean, healthy, and tasty, and it has become a popular food item in the United States.[12] To attract hunters, a few livestock men and women started breeding "big-rack bucks." Such breeding occurred more with whitetails than such other species as mule deer. The trendy idea developed to appeal to hunters who desired to land a so-called trophy buck.[13]

In response to such trends, livestock journals increasingly reported on such topics as "managing whitetails," "exotics," and "bison as a ranch income stream." By actual count, after 2000, there appeared, not counting weekly sales information, more feature articles on such topics as wildlife management, deer, and leasing of hunting land than on sheep, goats, wool, and mohair. A related decline occurred: In 1950, for example, stories on sheep dominated reporting in the *Livestock Weekly*, and nearly a half century later, in 1990, cattle led most of the journal's "news" articles.

Of course, in 2015, as it had for many years, *Livestock Weekly*, with Steve Kelton, its editor, included market updates on cattle, sheep, lambs, goats, wool, and mohair. It reviewed at some length legislative issues and labor regulations impacting agriculture. The April 30 issue contained an article suggesting a positive long-term outlook for cattle, which included C. A. Rodenberger's column "The Computer and the Cowboy". The magazine remains an important Southwestern farm and ranch publication.

A close cousin of *Livestock Weekly*, *Ranch and Rural Living*, is the official publication of the Texas Sheep and Goat Raisers Association. Its editor in 2015 was Gary Cutrer. The journal has evolved as the sheep and goat industry has changed. First called *The Texas Sheep and Goat Raisers' Magazine*, the publication became *Ranch Magazine* in the 1970s, reflecting the relative decline of sheep and goat operators vis-à-vis cattle producers. More recently it became *Ranch and Rural Living*, suggesting a growth in the number of its non-farm and ranch readers. Even so, the journal keeps its readers, and especially sheep, wool, goat, and mohair producers, informed on legislative matters, range land issues, water and conservation topics, breeding trends, product prices, marketing questions, and other important agricultural news. It is now in its ninety-fifth year of publication.

The Texas Sheep and Goat Raisers Association, the oldest sheep and goat organization in the United States, turned one hundred years old in 2015. Its initial purpose centered on the need for Texas wool and mohair growers to find a way to protect their flocks and herds from predators and thieves while also seeking important state legislation to aid the ovine husbandry. The group's service to the industry expanded during the subsequent century, and in recent years, the association adopted a broad program. In Washington, DC, and in Austin it aims to protect landowners' property rights, to fight what it calls "unnecessary government regulation," and to end the so called "death" tax—an inheritance issue. In a general sense, it also supports limits on national imports of goats and lambs, labelling and price reporting of imported lamb and goat meat, water rights for agricultural use, a permanent federal program for wool and mohair, and improved animal health regulations.[14]

In 2015, during their annual meeting in Kerrville, in the heart of the Texas Hill Country, sheep and goat raisers reaffirmed their organization's general goals. Also at the meeting, wool and lamb producers, buyers, processors, and their friends in such related groups as the American Boer Goat Association, Mohair Council of America, American Sheep Industry Association, and various auxiliaries celebrated the sheep and goat association's centennial, a truly grand milestone. Its headquarters, as it has for one hundred years, remains in San Angelo.

In 2015, Sandy Whittley served as executive secretary of the Texas Sheep and Goat Raisers Association. Energetic and dedicated, she has been with the organization since 1966. At their annual meeting in Kerrville, sheep and goat raisers recognized her in a special ceremony—a barbeque at the Roeder Ranch near Fredericksburg—for her remarkable half century of industry service. Besides overseeing the influential Texas group's activities, Whittley produced a column each month in *Ranch and Rural Living*. In it she wrote about agricultural issues and news, legislative matters, and marketing trends, and in general she has helped to keep members of the organization informed about the industry's current affairs.

As Sandy Whittley might attest, sheep and goat raising is a way-of-life in Texas, one with remarkable challenges but special rewards. It has supported growers in the state for more than two hundred years, providing not only food and fiber to a large population but also significant work in related occupations such as meat processing, woolen manufacturing, carpet production, agricultural sales and services, and transportation. For the foreseeable future, the Texas sheep and goat industry will continue to provide farmers and ranchers an oftentimes hard, a usually sustainable, and an always unique living.

1. Texas Sheep and Goat Raisers Association (hereafter TSGRA), www.tsgra.com, "Livestock."

2. Ibid. See also "Hair Sheep," *Ranch and Rural Living* 93, No. 10 (July 2012): 6.

3. *Texas Almanac, 2012–2013* (Denton, Texas: The Texas State Historical Association, 2012), 672, 686; *Texas Almanac, 2006–2007*, Sesquicentennial Edition, 1857–2007 (Dallas: The Dallas Morning News, L.P., 2007), 637–38, 640–41.

4. *Texas Almanac, 2006–2007*, 640–41.

5. *Texas Almanac, 2012–2013*, 105, 172, 683.

6. Ibid., 685, 687.

7. Ibid., 685.

8. Ibid., 671, 685, 687; TSGRA, www.tsgra.com, "Livestock."

9. "Renewal of Wool Act Apparent," *Livestock Weekly* (San Angelo) 37, No. 50 (December 19, 1985): 1.

10. "Edwards Aquifer Head Reviews Groundwater Status in Texas," *Livestock Weekly* 52, No. 25 (June 22, 2000): 6.

11. "$1.3 Billion Industry . . . ," *Ranch and Rural Living* 93, No. 6 (March 2012): 12.

12. "Bison as Ranch Income Stream," *Ranch and Rural Living* 93, No. 4 (January 2012) 25.

13. "Whitetail Deer Breeders—for Big Bucks," *Ranch and Rural Living* 93, No. 4 (January 2012): 20. See also "Alpacas," *Ranch and Rural Living* 93, No. 10 (July 2012): 5.

14. TSGRA, www.tsgra.com, "home."

APPENDIX

Classes of Sheep and Wool

WOOL classification was based on the condition, the quality, the strength and length of fiber, and the color of the fleece. Clothing wool included that produced by Merinos, Saxons, and other wools of Merino blood. Combing wool included that produced by Bakewell, Cotswold, Lincoln, and other like wools imported largely from England, and also hair of the alpaca, a cousin of the South American llama. Wool used in the manufacture of carpets and rugs was imported from Asia Minor (Turkey), Greece, Egypt, Syria, and other countries where native sheep had a coarse, wiry, tough fleece. While there is some disagreement on the point, most authorities maintained that very little carpet wool was produced in the United States in the nineteenth century.

For classifying wool as to "fineness" (diameter of fiber), two systems are used today: the "blood" (American) system and the "numerical" (English) system.

Most countries use the English system, which gauges the finest possible thread into which the wool can be spun. The fineness is measured in the number of fibers per inch.

American System	*English System*
Fine	80s
	70s
	64s
Half-Blood	60s
	58s
Three-Eighths Blood	56s
Quarter Blood	50s
	48s
Low-Quarter Blood	46s

Common	44s
Braid	40s
	36s

SOURCE: Vince Melone, President of the San Angelo Wool Processing Co., interview with Paul H. Carlson, April 28, 1977; Lamont Johnson, "Sheep Breeds 100 Years Ago," *Sheep and Goat Raisers' Magazine* 27 (June, 1947): 8–9, 12–13, 26–30.

Bibliography

Bibliographic Note

This bibliography is not an exhaustive listing of material on the Texas range sheep and goat industry. Rather, it is a listing of the sources cited in the notes plus a number of additional references. The standard works on the sheep industry in America, cited more completely below, are Wentworth, *America's Sheep Trails*, and Towne and Wentworth, *Shepherd's Empire*. *The Flock* by Austin, *Golden Fleece* by Call, *Sheep* by Gilfillan, and *The Golden Hoof* by Kupper are classics in insights, sympathies, and style. There has not been a detailed history of the Angora goat industry since 1900, when Black, *A New Industry—Or Raising the Angora Goat*, appeared.

The Southwest Collection at Texas Tech University, Lubbock, houses perhaps the most extensive records available for the Texas sheep and goat industry. The collection holds the large files of the Texas Sheep and Goat Raisers Association, the American Angora Goat Breeders Association, and the Mohair Council of America. It also holds oral history accounts from people engaged in the growing, transporting, processing, marketing, and manufacturing of wool and mohair, plus the personal papers of many people connected with the industry. It possesses nearly every book, article, and other source listed below.

Sources

ARCHIVAL COLLECTIONS

Austin. University of Texas, Archives. Earl Vandale Collection.

Boerne, Tex. Kendall County Abstract Co. Land Records.

Canyon, Tex. Panhandle-Plains Historical Museum. Archives.

Lubbock. Texas Tech University, Southwest Collection. American Angora Goat Breeders Association, Records. Black, William L., Papers. Jones, Marvin, Papers, 1909–76. Mohair Council of America, Records. Pierce, Victor I., Papers. Texas Sheep and Goat Raisers Association, Records. Thompson, Philip, Papers. Winslow Family Papers.

San Antonio. Institute of Texan Cultures. French File.

INTERVIEWS

Hubbs, Barry. Interview with Richie Cravens, July 8, 1977. Transcript in Southwest Collection, Texas Tech University, Lubbock.

Kelton, Elmer. Telephone interview with Paul H. Carlson, June 6, 1981. Transcript in possession of author.

Melone, Vince. Interview with Paul H. Carlson, April 28, 1977. Transcript in Southwest Collection, Texas Tech University, Lubbock.

Phillips, Hiram. Interview with Paul H. Carlson, October 1, 1981. Transcript in possession of author.

Posey, Walter. Interview with Seymour V. Connor, July 24, 1956. Transcript in Southwest Collection, Texas Tech University, Lubbock.

Richardson, Leo. Interview with Elmer Kelton, March 29, 1974. Transcript in Southwest Collection, Texas Tech University, Lubbock.

Sloan, T. P. Interview with Paul H. Carlson, August 23, 1981. Transcript in possession of author.

Swaim, E. L. Interview with Paul H. Carlson, March 29, 1980. Transcript in Southwest Collection, Texas Tech University, Lubbock.

Sloan, T. P. Interview with Paul H. Carlson, March 29, 1980. Transcript in possession of author.

Willoughby, Ray. Interview with Paul H. Carlson, April 28, 1977. Transcript in Southwest Collection, Texas Tech University, Lubbock.

THESES, DISSERTATIONS, AND MANUSCRIPTS

Bode, G. R. "The Life and Times of George Wilkins Kendall." Master's thesis, University of Texas, Austin, 1930.

Charles, Ralph. "Development of the Partido System in the New Mexico Sheep Industry." Master's thesis, University of New Mexico, 1940.

Evans, Samuel Lee. "Texas Agriculture, 1880–1930." Ph.D. dissertation, University of Texas, Austin, 1960.

Finck, Arthur L., Jr., trans. "The Regulated Emigration of the German Proletariat with Special Reference to Texas, by Dr. Von Herff." Master's thesis, University of Texas, Austin, 1949.

Fluth, Alice Freeman. "Indianola, Early Gateway to Texas." Master's thesis, St. Mary's University, 1939.

Gregg, John Ernest. "The History of Presidio County." Master's thesis, University of Texas, Austin, 1933.

Irving, M. P. "The Settlement of the Trans-Pecos or Big Bend Country." Master's thesis, University of Colorado, 1924.

Mills, H. J. Young. "A Study of the Wool Industry in Texas." Master's thesis, University of Texas, Austin, 1930.

Shearer, Ernest Charles. "Border Diplomatic Relations between the United States and Mexico, 1848–1860" Ph.D. dissertation, University of Texas, Austin, 1939.

Taylor, A. J. "*Pastores* in the Texas Panhandle." Photocopy in Southwest Collection, Texas Tech University, Lubbock.

Weaver, Bobby. "German Contributions to the Texas Cattle Industry." 1977 typescript. Copy in Southwest Collection, Texas Tech University, Lubbock.

NEWSPAPERS

Amarillo Sunday Globe-News, 1938.

Corpus Christi Caller, 1925–30.

Fort Worth Gazette, 1884.

Galveston News, 1880–84.

Houston Telegraph and Texas Register, 1841.

Marfa New Era, 1922.

San Angelo Standard-Times, 1954.

San Angelo Weekly Standard, 1895, 1937.

San Antonio Herald, 1859.

San Antonio Light, 1954.

Texas Livestock Journal (Fort Worth), 1891–97.

Wall Street Journal, 1978.

GOVERNMENT DOCUMENTS AND PUBLICATIONS

U.S. Congress, House. *Special Report on the History and Present Conditions of the Sheep Industry in the United States*. Prepared by E. A. Carman, H. A. Heath, and John Minto. House Misc. Doc. 105, 52 Cong., 2 Sess., 1892.

Gordon, Clarence. *Report on Cattle, Sheep and Swine, Supplementary to Enumeration of Livestock on Farms in 1880*. Washington, D.C.: Government Printing Office, 1880.

Jamieson, Stuart. *Labor Unionism in American Agriculture*. U.S. Department of Labor Bulletin 836, 1945. Reprint. New York: Arno Press, 1976.

Jones, John M., and J. E. Boog-Scott. *Sheep Production in Texas*. Bulletin of Texas Department of Agriculture. Austin, 1915.

Natural Fibers Economic Research. *Texas Sheep and Goat Wool and Mohair Industry*. Austin: Texas Department of Agriculture, 1974.

Texas Crop and Livestock Reporting Service. *Texas Historic Livestock Statistics, 1867–1976*. Bulletin 131. Austin: Texas Department of Agriculture, 1976.

———. *1975 Texas Livestock Statistics*. Bulletin 135. Austin: Texas Department of Agriculture, 1976.

U.S. Bureau of the Census. *Eleventh Census. Report of the Statistics of Agriculture, 1890*. Washington, D.C.: Government Printing Office, 1895.

———. *Tenth Census. Report on the Production of Agriculture, 1880*. Washington, D.C.: Government Printing Office, 1883.

U.S. Department of Agriculture. *Livestock on Farms, January 1, 1867–1919.* Bulletin of the Bureau of Agricultural Economics. Washington, D.C., 1938.

———. *Livestock, Meats, and Wool—Market Statistics and Related Data.* Bulletin of the Bureau of Agricultural Economics. Washington, D.C., 1942.

———. Bureau of Animal Husbandry. *Fourteenth Annual Report.* Washington, D.C.: Government Printing Office, 1891.

U.S. Department of Interior. *Eighth Census. Population, 1860.* Washington, D.C.: Government Printing Office, 1864.

Willingmyre, George T., et al. *The Angora Goat and Mohair Industry.* Misc. Circular No. 50, United States Department of Agriculture. Washington, D.C., 1929.

BOOKS

Austin, Mary. *The Flock.* Boston: Houghton Mifflin Co., 1914.

Barker, Eugene C., ed. *The Austin Papers.* 3 vols. Vol. 1, in two parts, published as Vol. 2 of the Annual Report of the American Historical Association for the Year 1919; Washington: Government Printing Office, 1924. Vol. 2 published as the Annual Report of the American Historical Association for the Year 1922; Washington: Government Printing Office, 1928. Vol. 3 published Austin: University of Texas, 1927.

Billington, Ray Allen. *Westward Expansion.* 4th ed. New York: Macmillan Co., 1974.

Black, William Leslie. *A New Industry—Or Raising the Angora Goat.* Fort Worth: Keystone Printing Co., 1900.

Bollaert, William. *William Bollaert's Texas.* Edited by W. Eugene Hollon and Ruth Lapham Butler. Norman: University of Oklahoma Press, 1965.

Bolton, Herbert Eugene. *Texas in the Middle Eighteenth Century.* Berkeley: University of California Press, 1915.

———, ed. *Athanase de Mézières and the Louisiana-Texas Frontier, 1768–1780.* 2 vols. Cleveland: Arthur H. Clark Company, 1914.

———, ed. *Spanish Exploration in the Southwest, 1542–1706.* New York: Charles Scribner's Sons, 1930.

Brown, Harry J., ed. *Letters from a Texas Sheep Ranch.* Urbana: University of Illinois Press, 1959.

Call, Hughie. *Golden Fleece.* Boston: Houghton Mifflin, 1942.

Castañeda, Carlos E. *Our Catholic Heritage in Texas, 1519–1936.* 7 vols. Austin: Von Boeckmann–Jones Co., 1936–58.

Clay, John. *My Life on the Range.* Chicago: Privately printed, 1924.

Clendenen, Clarence C. *Blood on the Border: The United States Army and the Mexican Irregulars.* London: Macmillan & Co., 1969.

Cole, Arthur Harrison. *The American Wool Manufacture.* 2 vols. New York: Harper & Row, Publishers, 1969.

Copeland, Fayette. *Kendall of the Picayune*. Norman: University of Okahoma Press, 1943.

Cox, James. *Historical and Biographical Record of the Cattle Industry and Cattlemen of Texas and Adjacent Territory*. St. Louis: Woodward and Tiernan Printing Co., 1895.

Faulk, Odie B. *A Successful Failure*. Austin: Steck-Vaughn Co., 1965.

Fenley, Florence. *Old Timers, Their Own Stories*. Uvalde: Hornby Press, 1939.

Forbis, William H. *The Cowboys*. New York: Time-Life Books, 1973.

Forrestal, P. P., trans. *The Solís Diary of 1767*. Vol. I, no. 6, Preliminary Studies of the Texas Catholic Historical Society. Austin, 1931.

Gilfillan, Archer. *Sheep*. Boston: Little, Brown & Co., 1929.

Grimm, Agnes G. *Llano Mestenas, Mustang Plains*. Waco: Texian Press, 1968.

Gulick, C. A.; Katherine Elliott; and Harriet Smither, eds. *The Papers of Mirabeau Bonaparte Lamar*. 5 vols. Austin: Pemberton Press, 1921–27.

Hafen, Leroy R.; W. Eugene Hollon; and Carl Coke Rister. *Western America*. 3d ed. Englewood Cliffs, N.J.: Prentice Hall, 1970.

Haley, J. Evetts. *Charles Goodnight: Cowman and Plainsman*. Boston: Houghton Mifflin Co., 1936.

Hill, Kate Adele. *Home Builders of West Texas*. San Antonio: Naylor Co., 1958.

History of the Cattlemen of Texas. Dallas: Johnston Printing and Advertising Co., 1914.

Hughes, Thomas, ed. *G.T.T.—Gone to Texas: Letters from our Boys*. London: Macmillan & Co., 1884.

Hughes, William Edgar. *The Journal of a Grandfather*. St. Louis: Privately printed, 1912.

Jordan, Terry G. *German Seed in Texas Soil: Immigrant Farmers in Nineteenth-Century Texas*. Austin: University of Texas Press, 1966.

Kenner, Charles L. *A History of New Mexico–Plains Indians Relations*. Norman: University of Oklahoma Press, 1969.

Klein, Julius. *The Mesta: A Study in Spanish Economic History*. Cambridge: Harvard University Press, 1920.

Kupper, Winifred. *The Golden Hoof: The Story of the Sheep of the Southwest*. New York: Alfred A. Knopf, 1945.

Latham, Hiram. *Trans-Missouri Stock Raising, the Pasture Lands of North America: Winter Grazing*. Edited by J. C. Dykes. Denver: Old West Publishing Co., 1962.

Lea, Tom. *The King Ranch*. 2 vols. Boston: Little, Brown & Co., 1957.

Lehmann, Val W. *Forgotten Legions, Sheep in the Rio Grande Plain of Texas*. El Paso: Texas Western Press, 1969.

Lewis, George M. *An Analysis of Shipments of Texas Sheep and Goats*. Bureau of Business Research. Monograph No. 7. University of Texas, Austin, 1930.

Linn, John J. *Reminiscences of Fifty Years in Texas*. New York: O. and J. Sadlier & Co., for the author, 1883.

McCarty, John L. *Maverick Town, the Site of Old Tascosa*. Norman: University of Oklahoma Press, 1946.

Madison, Virginia. *The Big Bend Country of Texas*. 2d rev. ed. New York: October House, 1968.

Mauersberger, H. A., ed. *Mathew's Textile Fibers*. 6th ed. London: Chapman & Hall, 1954.

Miller, Thomas Lloyd. *The Public Lands of Texas, 1519–1970*. Norman: University of Oklahoma Press, 1972.

Murrah, David J. *C. C. Slaughter: Rancher, Banker, Baptist*. Austin: University of Texas Press, 1981.

Myres, Sandra L. *The Ranch in Spanish Texas*. El Paso: Texas Western Press, 1969.

Nance, J. M. *After San Jacinto: The Texas-Mexican Frontier, 1836–1841*. Austin: University of Texas Press, 1963.

O'Connor, K. S. *The Presidio La Bahía del Espíritu Santo de Zuniga, 1721 to 1846*. Austin: Von Boeckmann–Jones Co., 1966.

Olmsted, Frederick Law. *A Journey Through Texas*. New York: Mason Brothers, 1860.

Parish, William J. *The Charles Ilfeld Company*. Cambridge: Harvard University Press, 1961.

Powers, Steven. *The American Merino: For Wool and for Mutton*. New York: Orange Judd Co., 1911.

Raht, Carlysle Graham. *The Romance of Davis Mountains and Big Bend Country*. El Paso: Rahtbooks Co., 1919.

Randall, Henry S. *Sheep Husbandry*. New York: Orange Judd Co., 1869.

Rathjen, Frederick W. *The Texas Panhandle Frontier*. Austin: University of Texas Press, 1973.

Richardson, Rupert N.; Ernest Wallace; and Adrian N. Anderson. *Texas, the Lone Star State*. 4th ed. Englewood Cliffs, N.J.: Prentice Hall, 1981.

Robertson, Pauline Durrett, and R. L. Robertson. *Panhandle Pilgrimage*. Canyon, Tex.: Staked Plains Press, 1973.

Tanner, Ogden. *The Ranchers*. Alexandria, Va.: Time-Life Books, 1977.

Taussig, Frank W. *The Tariff History of the United States*, 7th ed. New York: G. P. Putnam's Sons, 1923.

Texas Almanac and State Industrial Guide 1980–1981. Dallas: A. H. Belo Corporation, 1979.

Texas Almanac for 1858. Galveston: Richardson & Co., 1857.

Texas Almanac for 1859. Galveston: Richardson & Co., n.d.

Texas Almanac for 1868. Galveston: W. Richardson & Co., 1867.

Texas Almanac for 1861. Galveston: W. & D. Richardson, 1860.

Texas Almanac for 1867. Galveston: W. Richardson & Co., 1866.

Towne, C. W., and Edward N. Wentworth. *Shepherd's Empire*. Norman: University of Oklahoma Press, 1946.

Wallace, Ernest. *Texas in Turmoil.* Austin: Steck-Vaughn Co., 1965.

Webb, Walter P., and H. B. Carroll, eds. *The Handbook of Texas.* 2 vols. Austin: Texas State Historical Association, 1952.

Wentworth, Edward N. *America's Sheep Trials.* Ames: Iowa State College Press, 1948.

Williams, Amelia W., and Eugene C. Barker, eds. *The Writings of Sam Houston.* 8 vols. Austin: University of Texas, 1938–43.

Winslow, Edith Black. *In Those Days, Memories of the Edwards Plateau.* San Antonio: Naylor Co., 1950.

Woodman, Lyman L. *Cortina, Rogue of the Rio Grande.* San Antonio: Naylor Co., 1950.

Wright, Chester W. *Wool Growing and the Tariff: A Study in the Economic History of the United States.* Boston: Houghton Mifflin Co., 1910.

JOURNALS AND MAGAZINES

Agricultural Marketing Service. "Livestock Numbers in Texas." *Sheep and Goat Raiser* 22 (February, 1942): 12–16.

Archambeau, Ernest R. "The First Federal Census in the Panhandle, 1880." *Panhandle-Plains Historical Review* 23 (1950): 23–103.

Arnot, John. "My Recollections of Tascosa before and after the Coming of the Law." *Panhandle-Plains Historical Review* 6 (1933): 61–64.

Ashburn, Karl Everett. "Tariffs and Wool Duties since 1867." *Sheep and Goat Raisers' Magazine* 10 (November, 1929): 104–106.

Ashton, John. "The Golden Fleece of Early Days." *Sheep and Goat Raiser* 24 (December, 1943): 9–15.

———. "The Start of Sheep Breeding in Texas." *Sheep and Goat Raiser* 25 (December, 1944): 36–42.

Baker, T. Lindsay. "Turbine-Type Windmills of the Great Plains and Midwest." *Agricultural History* 54 (January, 1980): 38–51.

———. "Windmills of the Panhandle Plains." *Panhandle-Plains Historical Review* 53 (1980): 71–110.

"Birth of the National Wool Act." *National Wool Grower* 55 (January, 1965): 104–108.

Black, William L. "Ranching on 10¢ Land in Texas." *Cattleman* 14 (July, 1927): 31–32.

Blanton, W. E. "The Life of Captain Charles Schreiner Is an Inspiring Memory." *Sheep and Goat Raisers' Magazine* 7 (July, 1927): 9–11.

"Blazing the Way." *Sheep and Goat Raisers' Magazine* 7 (December, 1926): 4.

Brown, Harry J. "The Fleece and the Loom: Wool Growers and Wool Manufacturers during the Civil War Decade." *Business History Review* 29 (March, 1955): 1–27.

Campbell, Coleman M. "Era of Wool and Sheep in Nueces Valley." *Frontier Times* 2 (February, 1934): 195–96.

Carlson, Paul H. "Texas Background: Spanish or American?" *West Texas Historical Association Year Book* 52 (1976): 61–70.

Classen, John P. "History of Predatory Animal Work of the Bureau of Biological Survey in Texas." *Sheep and Goat Raisers' Magazine* 11 (June, 1931): 295.

Connor, Seymour V. "Early Ranching Operations in the Panhandle: A Report on the Agricultural Schedules of the 1880 Census." *Panhandle-Plains Historical Review* 27 (1954): 47–69.

"Corriedale Popularity Increases in Southwest." *Sheep and Goat Raiser* 22 (July, 1942): 33.

Davis, T. C. "The +IN Ranch: History and Development of a Pioneer Ranch." *Voice of the Mexican Border* 1 (October, 1933): 76–81.

De Leon, Arnoldo. "*Los Tosinques* and the Sheep Shearers Union of North America: A Strike in West Texas, 1934." *West Texas Historical Association Year Book* 55 (1979): 3–16.

Dugas, Vera L. "Texas Industry, 1860–1880." *Southwestern Historical Quarterly* 59 (October, 1955): 151–83.

Esplin, A.C. "The Rambouillet Breed Has a Future." *Sheep and Goat Raiser* 24 (October, 1943): 18.

Fenley, Florence. "Sheep Camps and Old-Time Mexican Herders." *Sheep and Goat Raiser* 25 (August, 1945): 8, 10–11.

Fenton, James I. "Big Spring's Amazing Tenderfoot: The Earl of Aylesford." *West Texas Historical Association Year Book* 55 (1979): 135–48.

———. "The Lobo Wolf: Beast of Waste and Desolation." *Panhandle-Plains Historical Review* 53 (1980): 57–70.

Gard, Wayne. "The Fence-Cutters." *Southwestern Historical Quarterly* 51 (July, 1947): 1–15.

Gipson, Fred. "Easy Money This Goat Money." *Sheep and Goat Raiser* 50 (October, 1969): 51, 59.

Gober, Harold M. "The Sheep Industry in Sterling County." *West Texas Historical Association Year Book* 27 (1951): 32–57.

Haley, J. Evetts. "Pastores del Palo Duro." *Southwest Review* 19 (April, 1934): 279–94.

Hamilton, T. R. "Trends in Production, Use and Prices of Texas Mohair." *Sheep and Goat Raiser* 25 (July, 1945): 14–17, 42–45.

———. "Trends in the Sheep Industry in Texas." *Sheep and Goat Raiser* 25 (June, 1945): 6–11, 46–47.

Harral, Arthur G. "Arthur G. Anderson, Pioneer Sheep Breeder of West Texas." *Southwestern Sheep and Goat Raiser* 6 (November, 1935): 12–13, 31.

Havens, T. R. "Livestock and Texas Law." *West Texas Historical Association Year Book* 36 (1960): 18–32.

———. "Sheepmen-Cattlemen Antagonism on the Texas Frontier." *West Texas Historical Association Year Book* 18 (1942): 10–23.

———. "Texas Sheep Boom." *West Texas Historical Association Year Book* 28 (1952): 3–17.

Hickman, Ervin. "Texas Has Twenty-five Million Dollar Angora Goat Industry." *Sheep and Goat Raiser* 16 (July, 1946): 20, 40–41.

"History of the National Wool Growers Association." *Southwestern Sheep and Goat Raiser* 7 (December, 1936): 28–30.

Hollon, W. Eugene. "Captain Charles Schreiner." *Southwestern Historical Quarterly* 47 (October, 1944): 145–68.

Holt, Roy D. "C. C. Doty, West Texas Pioneer." *Sheep and Goat Raiser* 22 (November, 1941): 20–21, 24–25, 28.

———. "Dominicker Hart Made Sheep Pay West of the Pecos." *Sheep and Goat Raiser* 30 (June, 1950): 49.

———. "Early-Day Stockmen on the Lower Trans-Pecos." *Cattleman* 24 (November, 1937): 11–14.

———. "Pioneer Cowmen of Brewster County and the Big Bend Area." *Cattleman* 29 (June, 1942): 13–16, 18–19, 21–29.

———. "Pioneering Sheepmen." *Sheep and Goat Raiser* 26 (December, 1945): 16, 18, 20–23.

———. "The Saga of Barbed Wire in Tom Green County." *West Texas Historical Association Year Book* 4 (1928): 32–49.

———. "Woes of the Texas Pioneer Sheepman." *Southwestern Sheep and Goat Raiser* 21 (December, 1940): 60–63.

Johnson, Lamont. "Sheep Breeds 100 Years Ago." *Sheep and Goat Raiser* 27 (June, 1947): 8–9, 12, 26–30.

Jones, J. M. "History of the Range Sheep Industry in Texas." *Southwestern Sheep and Goat Raiser* 6 (March, 1936): 8–9, 32, 34–35.

Kinchen, Oscar A. "Pioneers of No Man's Land." *West Texas Historical Association Year Book* 18 (1942): 22–26.

Kupper, Winifred. "Sheep Drive in the 90's." *Sheep and Goat Raiser* 26 (December, 1945): 40–43, 50.

———. "Folk Characters of the Sheep Industry." *Cattleman* 26 (March, 1940): 89–98, 100, 102–103.

Maxwell, Elizabeth Ann. "Experiences of the Eaheart Brothers, Sheep Traders." *West Texas Historical Association Year Book* 30 (1954): 157–61.

"Mohair Clip Sets New Record." *Sheep and Goat Raiser* 46 (March, 1966): 2B–3.

"New Era Dawns with Net-Wire Fences." *Sheep and Goat Raiser* 50 (October, 1969): 30, 38.

Noelke, H. C., Jr. "The History of the Corriedale Sheep." *Southwestern Sheep and Goat Raiser* 19 (August, 1939): 7–9.

"A Patriarch Passes to His Father." *Sheep and Goat Raisers' Magazine* 7 (February, 1927): 3.

Patterson, E. H. "Romance of the Rambouillet." *Sheep and Goat Raiser* 27 (October, 1946): 10–11.

Perry, Kenneth. "Indian Depredations in the Texas Big Bend: The Crosson Claims Case." *Panhandle-Plains Historical Review* 53 (1980): 35–56.

"Press Time Markets." *Ranch Magazine* 58 (June, 1978): 6.

Pritchell, V. M. "The Mohair Industry in the Southwest." *Monthly Business Review* 35 (December, 1950): 193–201.

"Rebuilding an Industry." *National Wool Grower* 55 (January, 1965): 99–103.

Romero, Jose Ynocencio. "Spanish Sheepmen on the Canadian at Old Tascosa." Edited by Ernest R. Archambeau. *Panhandle-Plains Historical Review* 19 (1946): 45–72.

"Saga of Colonel Black, The." *Sheep and Goat Raiser* 30 (July, 1950): 3–5.

"Sam Hill, Leader for Years, Dies at Tierra Alta." *Sheep and Goat Raisers' Magazine* 13 (July, 1933): 125.

"Sam Hill, Pioneer Builder of West Texas." *Sheep and Goat Raisers' Magazine* 7 (September, 1926): 8.

Scobee, Barry. "Highland Country Once Strictly Cattle, Now Vastly Changed." *Sheep and Goat Raiser* 22 (November, 1941): 11, 13.

Shaw, S. P. "Development of Wool and Mohair Warehouses in Texas." *Sheep and Goat Raisers' Magazine* 8 (May, 1928): 67–70.

Sutherland, W. G. "Adams Bros., Trans-Nueces Pioneers." *Cattleman* 17 (June, 1930): 17–21.

"T. A. Kincaid's Contribution to the Sheep and Goat Industry." *Sheep and Goat Raiser* 30 (January 1950): 20–23.

Taylor, A. J. "*Pastores* in the Texas Panhandle." Edited by Robert E. Simmons. *Hale County History* 10 (February, 1980): 1–46.

Utley, Robert M. "The Range Cattle Industry in the Big Bend of Texas." *Southwestern Historical Quarterly* 69 (April, 1966): 417–41.

Wentworth, Edward N. "The Evolution of Sheep Shearing in America." *Sheep and Goat Raiser* 25 (June, 1945): 22, 24–27, 30, 32.

———. "The Golden Fleece in Texas." *Sheep and Goat Raiser* 24 (December, 1943): 16–17, 20, 25–33.

———. "Sheep Trails of Early Texas (part 1)." *Southwest Sheep and Goat Raiser* 9 (June, 1939): 7, 35, 37, 39.

———. "Sheep Trails of Early Texas (part 2)." *Southwestern Sheep and Goat Raiser* 9 (July, 1939): 23, 34.

"Wolf Proof Fences in Texas." *Cattleman* 18 (September, 1931): 27–28.

"World War I's Impact on Wool and Lamb." *National Wool Grower* 55 (January, 1965): 46–48.

Young, Irene, et al. "Men, Sheep and 100 Years." *National Wool Grower* 55 (January, 1965): 1–105.

Index